T0137209

Intelligent Systems Reference Library

Volume 191

The aim of this series is to publish a Reference Library, including novel advances and developments in all aspects of Intelligent Systems in an easily accessible and well structured form. The series includes reference works, handbooks, compendia, textbooks, well-structured monographs, dictionaries, and encyclopedias. It contains well integrated knowledge and current information in the field of Intelligent Systems. The series covers the theory, applications, and design methods of Intelligent Systems. Virtually all disciplines such as engineering, computer science, avionics, business, e-commerce, environment, healthcare, physics and life science are included. The list of topics spans all the areas of modern intelligent systems such as: Ambient intelligence, Computational intelligence, Social intelligence, Computational neuroscience, Artificial life, Virtual society, Cognitive systems, DNA and immunity-based systems, e-Learning and teaching, Human-centred computing and Machine ethics, Intelligent control, Intelligent data analysis, Knowledge-based paradigms, Knowledge management, Intelligent agents, Intelligent decision making, Intelligent network security, Interactive entertainment, Learning paradigms, Recommender systems, Robotics and Mechatronics including human-machine teaming, Self-organizing and adaptive systems, Soft computing including Neural systems, Fuzzy systems, Evolutionary computing and the Fusion of these paradigms, Perception and Vision, Web intelligence and Multimedia.

** Indexing: The books of this series are submitted to ISI Web of Science, SCOPUS, DBLP and Springerlink.

More information about this series at http://www.springer.com/series/8578

Natalia A. Serdyukova · Vladimir I. Serdyukov

Algebraic Identification of Smart Systems

Theory and Practice

 Springer

Natalia A. Serdyukova
Plekhanov Russian University
of Economics
Moscow, Russia

Vladimir I. Serdyukov
Department of Applied Mathematics
Bauman Moscow State Technical University
Moscow, Russia

ISSN 1868-4394   ISSN 1868-4408 (electronic)
Intelligent Systems Reference Library
ISBN 978-3-030-54472-0   ISBN 978-3-030-54470-6 (eBook)
https://doi.org/10.1007/978-3-030-54470-6

This Springer imprint is published by the registered company Springer Nature Switzerland AG
The registered company address is: Gewerbestrasse 11, 6330 Cham, Switzerland

Foreword

The book *Algebraic Identification of Smart Systems. Theory and Practice* continue the book *Algebraic Formalization of Smart Systems. Theory and Practice*.

The following main results were obtained in it: a method for studying intelligent systems using the new concept of a quasi-fractal algebraic system, a new concept of a structurally stable quasi-fractal intellectual system, a theorem explaining the occurrence of mutations, the concept of a quasi-fractal scale and level of measurability. Two models of algebraic formalization of the knowledge system representation are constructed: in the form of a free group of factors that determine the knowledge system, and in the form of a semantic network, which is represented by a finite Boolean algebra, the concept of parametric algebraic potential of a smart system is introduced. The questions of randomness of the system structure, system infrastructure, stability and integrity of the system are investigated. The concept of probability-isomorphic groups is introduced. The concept of the Erdös–Renyi algorithm on finite graphs and the concept of the Cayley graph of a group are used for this purpose. It has been shown that the problem of smart system identification is extremely important when studying mathematical objects. The concept of smart infrastructure is considered from the point of view of tensor estimation. The concept of structural stability of a closed associative smart system with feedback is introduced and considered. Theorems on tensor estimates of structurally stable systems are obtained.

The main tool is the notion of a quasi-fractal algebraic system and the theory of *P*—purities of algebraic systems. In 1990s, the concept of purities by predicates was introduced by one of the authors and later on authors found out some applications of the theory of purities by predicates to practice.

The main methodology is the methodology of the theory of algebraic systems, as a synthesis of algebra and logic, discovered by A. I. Maltsev, and a generalization of the concept of an algebraic system to the level of fractals—quasi-fractal algebraic systems introduced in this book, based on the theory of fractals developed by Benoit Mandelbrot.

Basic methods are: fixed-point theorem, the use of well-known theorems of group theory and the theory of Boolean algebras, Erdös–Renyi algorithms.

The book which is offering to you, *The Algebraic Identification of Smart Systems. Theory and Practice* is an attempt to use the opportunity to investigate a smart system in more detail, adding new factors to the model of factors determining the system that describe previously indivisible elements of the initial model of factors determining the system, that is an attempt to use zoom.

In view of the sufficient complexity of the issues under consideration, we dwell briefly on the contents of the chapters.

In Chap. 1, the problem of identification of a smart system is considered from the standpoint of compliance of the functioning of the system with its purpose of functioning and evaluation of this compliance. Here, authors construct a structural (by the structure of system attributes) analogue of the econometric method of principal components.

Main idea of Chap. 2, concerning the problem of representation of a model of a system, runs as follows: Authors would like to find such a form of a model for a smart system that one can refine its properties. The quasi-fractal systems allow one to use the concept of algebraic formalization of smart systems for a more in-depth study of smart systems, using the possibilities of its study, being at different levels of smart system's detail.

In Chap. 3, authors describe system's functions from the point of view of quasi-fractal algebraic systems. Also, natural fractal (algebraic) systems are introduced and authors show that they can be defined in the form of quasi-fractal algebraic systems. Writing natural fractals in the form of a quasi-fractal algebraic system allows one to represent the process of decomposition of the system when moving from the upper levels of the quasi-fractal algebraic system to the lower ones and, thus, allows to describe more clearly the functions of the system, that is to use zoom.

In Chap. 4, measuring processes in complex systems modeled by quasi-fractal algebraic systems are discussed. The notion of a quasi-fractal scale is defined that allows to monitor and diagnostics in complex systems modeled by quasi-fractal algebraic systems. The level of measurability by a quasi-fractal scale is determined using the Brouwer fixed-point theorem.

In Chap. 5, authors consider the test system as a measurement system through assessments of students' knowledge. The main issues here are the question of assessing the adequacy of the measurement results to the real level of knowledge, skills of students. In this regard, two models of the algebraic formalization of the representation of the knowledge system are considered: in the form of a free group of factors that determine the knowledge system and in the form of a semantic network, which is represented using finite Boolean algebra. These models allow one to show that all binary tests, i.e., test involving answers in the form of either "yes" or "no," can be solved true, without knowing the specific nature of the proposed questions.

Analogues of physical concepts and, in particular, the concept of potential are now used in various fields of knowledge related to the humanities, such as, for example, economics, teaching theory, control theory and so on. In Chap. 6, authors introduce the concept of parametric algebraic potential in line with the concept of algebraic formalization of the system. Let us remind that the potential energy (in our context is a measure of the transition of a system from one state to another is determined by the mutual position of interacting bodies or parts of the same body (subsystems in our context), that is, of the structure of the system. Here, the notions of Borel function and Borel measure are used.

The main questions of Chap. 7 are: the randomness of the structure of the system, the infrastructure of the system, its sustainability and the integrity of the system. The following basic questions are considered—random graphs, analogues of the Erdös–Renyi model for quasi-fractal graphs, quasi-fractal homomorphisms, scales, measures (continuation of Chap. 4), monitoring the functioning of a system represented by a quasi-fractal random graph, system structure randomness and integrity, infrastructure, giant component, the notion of random quasi-fractal graph (continuation of Chap. 4), random quasi-fractal scales (continuation of Chap. 4).

In Chap. 8, the concept of probability-isomorphic groups is introduced. To do this, the concept Erdös–Renyi algorithm on finite graphs and the concept of a Cayley graph of a group are used. Some notions of category theory are considered here because many algebraic notions are in deep connection with it: Saunders Mac Lane [Saunders Mac Lane "Categories for the Working Mathematicians, Second Edition, Springer-Verlag New York, Inc, 1998, 315 pp."] has shown that many algebraic objects, such as a group, a monoid, a field of fractions of a domain of integrity, etc., can be represented as a category. On the other hand, each category can be considered as a graph. The study of these issues has shown that the problem of identifying a system is extremely important in the study of mathematical objects.

In Chap. 9, the concept of smart infrastructure is examined from the standpoint of tensor estimation. The main questions of Chap. 9 are the following ones—verbal tensor estimates in the smart systems theory, methods to increase system reliability, tensor estimation of system quality, the algorithm of tensor estimation of system quality, formalization the approaches and the concept of system sustainability, the algorithm of tensor estimation of system sustainability, sustainability of a quasi-fractal system (a system modeled by a quasi-fractal algebraic system) and system sustainability violation: violation of the system's closeness property; malfunction of one of the system's factors, violation of feedback in the system; factor-flexible quasi-fractal system, substitution of system's functions; system's compensational possibilities, compensational functions of a factor-flexible quasi-fractal system.

In Chap. 10, smart systems modeled by solvable groups and quasi-fractal systems which modeled by quasi-fractal solvable groups are considered. In this chapter, the concept of structural sustainability of a closed associative system with a feedback is introduced and considered. It is shown that closed innovation systems with commuting determining factors do not have the property of structural sustainability and a closed associative system with a feedback modeled by a free nonabelian group is structurally sustainable with a minimum regulation range. Theorems about tensor estimates of structurally sustainable groups are obtained.

The book is intended for all interested in these issues.

Professor Vladimir A. Bubnov
Doctor of Technical Sciences
Institute of Digital Education
Moscow City University
Moscow, Russia

Seregina Svetlana
Professor
Doctor of Sciences in Economics

Deputy Head of Department of
Theoretical Economics

Faculty of Economic Sciences
National Research University Higher
School of Economics
Moscow, Russia

Preface

The book *Algebraic Identification of Smart Systems. Theory and Practice* is in fact a second part of the book *Algebraic Formalization of Smart Systems. Theory and Practice*.

We have considered in it the new concept of quasi-fractal algebraic system based on the A. I. Maltsev's theory of algebraic systems on the theory of fractals developed by Benoit Mandelbrot to investigate a smart system in more detail, adding new factors to the model of factors determining the system that describe previously indivisible elements of the initial model of factors determining the system.

This concept allows us to obtain the following main results: a method for studying intelligent smart systems, a new concept of a structurally stable quasi-fractal intellectual smart system, the concept of parametric algebraic potential of a smart system, the concept of a quasi-fractal scale and level of measurability of a smart system, an explanation the occurrence of mutations. Main kinds of algebraic systems we used for it are groups and Boolean algebras. Also, methods of probability theory have been used. Randomness of the system structure, system infrastructure, stability and integrity of the smart system are investigated from this point of view. The concept of the Erdös–Renyi algorithm on finite graphs and the concept of the Cayley graph of a group are used. The concept of probability-isomorphic groups is introduced. The book contains examples: application of the results obtained to the knowledge system and to issues of economics and finance. Two models of algebraic formalization of the knowledge system representation are constructed: in the form of a free group of factors that determine the knowledge system and in the form of a semantic network, which is represented by a finite Boolean algebra. The notion of smart infrastructure is considered from the point of view of tensor estimation. Structural stability of a closed associative smart system with a feedback is introduced and considered. Theorems on tensor estimates of structurally stable systems are obtained.

The main tool we used here, quasi-fractal algebraic systems, helps us to see a smart system in more details by adding new factors to the model of factors determining the system to describe previously indivisible elements of the initial model that is as to use scaling and zoom.

Basic methods are: fixed-point theorem, the use of well-known theorems of group theory and the theory of Boolean algebras, Erdös–Renyi algorithms.

Here, we give brief content of the book to clear up book logic, all connections between its chapters, main notions and results. The numeration coincides with the one we use in the book.

Chapter 1. Smart Systems Identification Problem

Abstract.

The main issues here are:

In the first Chapter, the problem of identification of a smart system is considered from the standpoint of compliance of the functioning of the system with its purpose of functioning and evaluation of this compliance. The problem of identification of smart systems is to answer the question: does the system satisfy the objective of its operation at each moment in time, and if so, how much? Thus, the problem of identification is divided into the following two tasks:

1. Does the system S satisfy the objective of its operation $P(S)$?
2. To assess how the system S satisfies the goals of its functioning $P(S)$?

Also, we construct a structural (by the structure of system attributes) analogue of the econometric method of principal components, developed by the outstanding Russian algebraist D. K. Faddeev.

Keywords: Identification problem · Smart system · Identification matrix

1.1 Introduction

Main idea of Chap. 1 concerning the problem of system identification runs as follows:

The purpose of the functioning of the economic system is to become a smart system.

The purpose of the functioning of the financial system is to become a smart system.

The purpose of the functioning of the training system is to become a smart system. We distinguish classes of systems whose purpose of functioning is to become a smart system, or come to the final state of smart.

1.2 The Problem Statement. Problems of Identification and Pattern Recognition: What is the Difference

1.2.1 Identification Problem Formalization

1.2.2 Possible Approaches to Solve Identification Problem. System Identification up to Internal or External Attributes

The idea of solving the problem defined by the title of this section is to determine the equivalence relation on the set of predicates and then use the analogy of the introduced concepts with the notion of quasi-isomorphism of Abelian groups.

Here we define a probabilistic space of internal attributes of the system S, a probabilistic space of external attributes of the system S.

1.3 System Description up to External Attributes. (Systems Classification up to External Attributes)

1.3.1 Examples. Test System for Knowledge Control. Connection of System Identification Problem with Classical Logic

Example 1 Test system for knowledge control.
Example 2 Connection of system identification problem with classical logic.

1.4 System's Identification Matrix as an Analogue of the Characteristic Function

Here we construct an analogue of the characteristic function—the system's identification matrix.

1.5 Systems with Two Subsystems and Two Internal Attributes

1.5.1 Orthogonality. The Subsystem S_j Identifier in the System S. Structural Analogue of the Method of Principal Components

In this section, we construct a structural (by the structure of system attributes) analogue of the econometric method of principal components, developed by the outstanding Russian algebraist D. K. Faddeev.

1.6 Decomposition of the System by its Basic Subsystems and Property. Decomposition of Attributes of a System by its Basic Attributes and a Subsystem

Here we introduce the definitions that allow one to decompose the system by the selected property of the system and by its subsystems, and by its selected subsystem and a fixed property.

1.7 Probability Identification Matrices

1.8 Examples of System Models with the Same Set of Basis Subsystems and Defining Attributes, the Same Matrix $M = ||F(Q_i(S_j))||$ and Different Structures

Chapter 2. The Complexity of the Identification Problem. Transition from by Element-wise Model of the System to the Model of Factors Determining the System. Quasi-fractal Algebraic Systems

Abstract. In Chap. 2 the following basic questions are considered:

- models' presentation forms,
- statistics methods of system identification,
- algebraic methods of systems identification,
- transition from the element-wise model of the system to the model of factors determining the system,
- fractal forms of model representation and systems identification methods,
- quasi-fractal algebraic systems.

Keywords: Identification problem · Quasi-fractal system · Contraction mapping · Elementary theory

2.1 Introduction

Main idea of Chap. 2 concerning the problem of representation of a model of a system, runs as follows: we would like to find such a form of a model for a smart system that one can refine its properties. The quasi-fractal systems allow one to use the concept of algebraic formalization of smart systems for a more in-depth study of smart systems, using the possibilities of its study, being at different levels of smart system's detail.

Advantages of the algebraic quasi-fractal form of representation of models are as follows.

The quasi-fractal form of representing the models of the system S in the algebraic formalization of systems allows us to consider major, initially distinguished factors that determine the system, in the form of algebraic systems, that is, it allows, in fact, to refine the model G_S of factors determining the system S, onto any level $k = 1, \ldots n, \ldots$

2.2 Models' Presentation Forms

2.3 Statistics Methods of System Identification

2.4 Algebraic Methods of Systems Identification. Transition from the Element-wise Model of the System to the Model of Factors Determining the System

2.5 Fractal Forms of Model Representation and Systems Identification Methods

2.6 Quasi-Fractal Algebraic Systems

Definition 2.1 Let's consider the algebraic system $A_1 = \langle A_1; \Omega_1 \rangle$ of the signature Ω_1, such that every element $a_\alpha, \alpha \in \Lambda_1$, of the main set A_1 of the system A_1 in turn is an algebraic system of the signature Ω_2. That is $a_\alpha = A_2^\alpha = \langle A_2^\alpha; \Omega_2 \rangle, \alpha \in \Lambda_1 s$ an algebraic system of the second level. Continue this process by induction. If an algebraic system $A_k^\alpha = \langle A_k^\alpha; \Omega_k \rangle, \alpha \in \Lambda_k$ is an algebraic system of the level k of the fractal and every element $a_\alpha, \alpha \in \Lambda_k$ of the main set A_k^α of the system A_k^α is an algebraic system $a_\alpha = A_{k+1}^\alpha = \langle A_{k+1}^\alpha; \Omega_{k+1} \rangle$ of the signature Ω_{k+1}, of the level $k+1$, $\alpha \in \Lambda_{k+1}$, of the fractal, then the algebraic system $A_1 = \langle A_1; \Omega_1 \rangle$ is called a quasi-fractal algebraic system. If all signatures $\Omega_k, k = 1, \ldots n, \ldots$, are equal to each other and all the systems $\langle A_k^\alpha; \Omega_k \rangle, \alpha \in \Lambda_k$, are isomorphic to each other then the algebraic system $A_1 = A_1; \Omega_1$ is called a fractal algebraic of the signature Ω_1.

We should explain that in this notation $A_1 = \langle A_1; \Omega_1 \rangle$ plays the role of a universal variable to denote a quasi-fractal algebraic system.

Let's note that an ordinary algebraic system is a quasi-fractal system of the first level.

2.6.1 Contraction Mappings of a Quasi-fractal Algebraic System

Fractal mathematical models are characterized by compression processes. In fact, in algebraic quasi-fractals there is a process of "qualitative compression" when moving across the fractal levels from top to bottom. At the same time, one can also build contraction quasi-fractal mappings and compress quasi-fractal levels in different ways. Here we consider some examples.

2.6.2 Metric Spaces. Contraction Mappings Examples

Key point here is

Brouwer's Fixed-Point Theorem

If $f: X \to X$ is a continuous map of a convex compact X into itself, then there exists a fixed point of the map f.

The main theorem here is the following one.

Main Theorem

Let the system S be modeled by a quasi-fractal algebraic system $A_1 = \langle A_1; \Omega_1 \rangle$. Then the system S has a finite level of forecasting (predictability).

2.7 Algebraic Fractals

In this section, we consider questions about how, using an algebraic fractal system, that is, using an algebraic quasi-fractal, one can write the following well-known constructions of algebra and category theory:

– free product,
– Cartesian product,
– direct sum.

2.7.1 Product and Coproduct Record of Category Theory in the Form of a Quasi-fractal Algebraic System

2.7.2 Record (Cartesian) Product of a Countable Number of Algebraic Systems in the Form of a Quasi-fractal Algebraic System (in the Form of an Algebraic Fractal)

2.7.3 Record Coproduct of a Countable Number of Algebraic Systems in the Form of a Quasi-fractal Algebraic System (in the Form of an Algebraic Fractal)

2.7.4 Special Case. Quasi-fractal Groups

2.8 Quasi-fractal Graphs

Definition   2.11 Let's   consider   the   graph $\Gamma_1 = \langle \{V^1_{\alpha_1} | \alpha_1 \in \Lambda_1 \},$
$\{u^1_{\alpha_1,\beta_1} | \alpha_1, \beta_1 \in \Lambda_1 \} \rangle$, where $\{V^1_{\alpha_1} | \alpha_1 \in \Lambda_1 \}$ is the set of vertices of the graph Γ_1,
$\{u^1_{\alpha_1,\beta_1} | \alpha_1, \beta_1 \in \Lambda_1 \}$ is the set of edges of the graph Γ_1.

Let's call the graph Γ_1 a graph of the first level. Let, in turn, $V^1_{\alpha_1}$ be the graph $\Gamma_2 = \langle \{V^2_{\alpha_2} | \alpha_2 \in \Lambda_2 \}, \{u^2_{\alpha_2,\beta_2} | \alpha_2, \beta_2 \in \Lambda_2 \} \rangle$, where $\{V^2_{\alpha_2} | \alpha_2 \in \Lambda_2 \}$ is the set of vertices of the graph Γ_2, $\{u^2_{\alpha_2,\beta_2} | \alpha_2, \beta_2 \in \Lambda_2 \}$ is the set of edges of the graph Γ_2.

Let's call the graph Γ_2 a graph of the second level. Let's continue this process by induction.

If the graph $\Gamma_k = \left\langle \left\{ V^k_{\alpha_k} | \alpha_k \in \Lambda_k \right\}, \left\{ u^k_{\alpha_k, \beta_k} | \alpha_k, \beta_k \in \Lambda_k \right\} \right\rangle$, where $\left\{ V^k_{\alpha_k} | \alpha_k \in \Lambda_k \right\}$ is the set of vertices of the graph Γ_k, $\left\{ u^k_{\alpha_k, \beta_k} | \alpha_k, \beta_k \in \Lambda_k \right\}$ is the set of edges of the graph—Γ_k, is the graph of the level k, then let $V^k_{\alpha_k}, \alpha_k \in \Lambda_k$, be a graph $\Gamma_{k+1} = \left\langle \left\{ V^{k+1}_{\alpha_{k+1}} | \alpha_{k+1} \in \Lambda_{k+1} \right\}, \left\{ u^{k+1}_{\alpha_{k+1}, \beta_{k+1}} | \alpha_{k+1}, \beta_{k+1} \in \Lambda_{k+1} \right\} \right\rangle$, where $\left\{ V^{k+1}_{\alpha_{k+1}} | \alpha_{k+1} \in \Lambda_{k+1} \right\}$ is a set of vertices of the graph Γ_{k+1}, $\left\{ u^{k+1}_{\alpha_{k+1}, \beta_{k+1}} | \alpha_{k+1}, \beta_{k+1} \in \Lambda_{k+1} \right\}$ is the set of edges of the graph Γ_{k+1}. The graph Γ_{k+1} is called a graph of the level $k+1$. In this case the graph $\Gamma_1 = \left\langle \left\{ V^1_{\alpha_1} | \alpha_1 \in \Lambda_1 \right\}, \left\{ u^1_{\alpha_1, \beta_1} | \alpha_1, \beta_1 \in \Lambda_1 \right\} \right\rangle$ is called a quasi-fractal graph. If all the graphs $\Gamma_k = \left\langle \left\{ V^k_{\alpha_k} | \alpha_k \in \Lambda_k \right\}, \left\{ u^k_{\alpha_k, \beta_k} | \alpha_k, \beta_k \in \Lambda_k \right\} \right\rangle, r = 1, \ldots, n, \ldots$ are isomorphic to each other, then the graph $\Gamma_1 = \left\langle \left\{ V^1_{\alpha_1} | \alpha_1 \in \Lambda_1 \right\}, \left\{ u^1_{\alpha_1, \beta_1} | \alpha_1, \beta_1 \in \Lambda_1 \right\} \right\rangle$ is called a fractal graph. The graph $\Gamma_k = \left\langle \left\{ V^k_{\alpha_k} | \alpha_k \in \Lambda_k \right\}, \left\{ u^k_{\alpha_k, \beta_k} | \alpha_k, \beta_k \in \Lambda_k \right\} \right\rangle$, where $\left\{ V^k_{\alpha_k} | \alpha_k \in \Lambda_k \right\}$ is the set of vertices of the graph Γ_k, $\left\{ u^k_{\alpha_k, \beta_k} | \alpha_k, \beta_k \in \Lambda_k \right\}$ is the set of edges of the graph Γ_k, is called a quasi-fractal graph of the level k.

The notion of a quasi-fractal graph is a generalization of the notion of a graph, since an ordinary graph is a quasi-fractal graph of the first level.

Examples An example of a fractal graph without edges is the Dirichlet function:

$$D(x) = \left\{ \begin{array}{l} 1, x \in Q \\ 0, x \in R \backslash Q \end{array} \right.$$

It is well known that $D(x) = \lim_{m \to \infty} \lim_{n \to \infty} cos^{2n}(m! \pi x)$. The Dirichlet function is periodic one with a period equal to any rational number:

$D(x) = D(x+q)$ for any $q \in Q$.

2.8.1 Decomposition and Synthesis Based on Quasi-fractal System Models

2.8.2 Appendix. Quasi-fractal Systems Recognition Methods. Monitoring Learning Outcomes Through Testing

2.9 Example. The Control Algorithm of a Complex System Based on a Quasi-fractal Model

2.9.1 Elementary Controlled Systems

The question arises: Can an algorithm to achieve the goal of the system be built?

Definition 2.13 A system S is called elementary controllable if the elementary theory of its model G_S is solvable.

Examples

1. Semantic networks represented by finite Boolean algebras.
2. Systems modeled by Abelian groups. In particular, the system from point 2.8, example, (which is a fractal cyclic group $Z_2 = \langle Z_2| +, -, 0\rangle$) is elementary controllable.

2.10 Conclusion. Connection of Quasi-fractals with Synergetics

Chapter 3. General System Function

> Large-scale invariance, or its self-similarity of
> fractal structure, is its
> characteristic property
> F. A. Tzitsin, Fractal Universe,
> Home page http://www.delphis.ru/journal/article/fraktalnaya-
> vselennaya

Abstract. In Chap. 3 the following basic questions are considered:

- natural fractal algebraic systems through series of examples from mathematics, natural sciences, telecommunications,
- the concept of quasi-fractal homomorphism and target quasi-fractal of a system, system target functions,
- regulation of a quasi-fractal system. Regulatory functions of the system. Theorems explaining the appearance of mutations.

Keywords: Testing · Binary tests · System efficiency · Knowledge system · Semantic network · Probability measure

3.1 Introduction

In Chap. 3 we shall approach the description of system functions from the point of view of quasi-fractal algebraic systems introduced in Chap. 2. Before doing this, we define natural fractal (algebraic) systems and show that they can be defined in the form of quasi-fractal algebraic systems. Writing natural fractals in the form of a quasi-fractal algebraic system allows one to represent the process of decomposition of the system when moving from the upper levels of the quasi-fractal algebraic system to the lower ones, and, thus, allows to describe more clearly the functions of the system.

3.2 Natural Fractal Algebraic Systems

We begin with preliminary remarks and some examples of natural fractals.

Definition 3.1 A quasi-fractal is called natural if each of its elements is a subsystem of the original system under study.

In the common case it is not so, see Chap. 2, Definition 2.1.

Definition 2.1, Chap. 2 Let's consider the algebraic system $A_1 = \langle A_1; \Omega_1 \rangle$ of the signature Ω_1, such that every element $a_\alpha, \alpha \in \Lambda_1$, of the main set A_1 of the system A_1 in turn is an algebraic system of the signature Ω_2. That is $a_\alpha = A_2^\alpha = \langle A_2^\alpha; \Omega_2 \rangle, \alpha \in \Lambda_1$ is an algebraic system of the second level. Continue this process by induction. If an algebraic system $A_k^\alpha = \langle A_k^\alpha; \Omega_k \rangle, \alpha \in \Lambda_k$ is an algebraic system of the level k of the fractal and every element $a_\alpha, \alpha \in \Lambda_k$ of the main set A_k^α of the system A_k^α is an algebraic system $a_\alpha = A_{k+1}^\alpha = \langle A_{k+1}^\alpha; \Omega_{k+1} \rangle$ of the signature Ω_{k+1}, of the level $k+1$, $\alpha \in \Lambda_{k+1}$, of the fractal, then the algebraic system $A_1 = \langle A_1; \Omega_1 \rangle$ is called a quasi-fractal algebraic system. If all signatures $\Omega_k, k = 1, \ldots n, \ldots$, are equal to each other and all the systems $\langle A_k^\alpha; \Omega_k \rangle, \alpha \in \Lambda_k$, are isomorphic to each other then the algebraic system $A_1 = \langle A_1; \Omega_1 \rangle$ is called a fractal algebraic of the signature Ω_1.

We should explain that in this notation $A_1 = \langle A_1; \Omega_1 \rangle$ plays the role of a universal variable to denote a quasi-fractal algebraic system.

Let's note that an ordinary algebraic system is a quasi-fractal system of the first level.

3.3 The Concept of Quasi-fractal Homomorphism. Target Quasi-fractal of a System. System Target Functions

Definition 3.2 Let $A_1 = \langle A_1; \Omega_1 \rangle, B_1 = \langle B_1; \Omega_1 \rangle$ be quasi-fractal algebraic systems such that for each level of a quasi-fractal systems the systems $A_k^\alpha = \langle A_k^\alpha; \Omega_k \rangle$ and $B_k^\alpha = \langle B_k^\alpha; \Omega_k \rangle$ have one the same signature Ω_k. A mapping $f : A_1 = \langle A_1; \Omega_1 \rangle \to B_1 = \langle B_1; \Omega_1 \rangle$ is called a quasi-fractal homomorphism from a system $A_1 = \langle A_1; \Omega_1 \rangle$ into a system $B_1 = \langle B_1; \Omega_1 \rangle$, if the following two conditions are fulfilled:

(1) For every $\alpha \in \Lambda_k$ the equality $f(a_\alpha) = f(b_\alpha)$ holds for some $b_\alpha \in \Lambda_k$, that is f maps the elements of the quasi-fractal algebraic system $A_1 = \langle A_1; \Omega_1 \rangle$ of the level $k+1$ into elements of the same level $k+1$ of the quasi-fractal algebraic system $B_1 = \langle B_1; \Omega_1 \rangle$;
(2) f saves all operations and predicates of the signature Ω_k. for all levels k of the quasi-fractal. That is: $f(F_\xi(x_1, \ldots, x_{m_\xi})) = (f(F_\xi))(f(x_1), \ldots, f(x_{m_\xi}))$ for every operation $F_\xi \in \Omega_k$ and every $x_1, \ldots, x_{m_\xi} \in A_k^\alpha$ and $P_\eta((x_1, \ldots, x_{n_\eta})) \Leftrightarrow (f(P_\eta))(f(x_1), \ldots, f(x_{n_\eta}))$ for every predicate $P_\eta \in \Omega_k$ and every $x_1, \ldots, x_{n_\eta} \in A_k^\alpha$.

3.3.1 Algorithm of Construction a Level of a System Predictability

1. Next, according to the algorithm described in Chap. 2, Examples 1, 2, 3, we construct a contraction map of a quasi-fractal algebraic system $Hom_1(G_S, G_S)$. After that we define the metric d on the quasi-fractal $Hom_1(G_S, G_S)$ in accordance with Example 4, Chap. 2.

   We embed the metric quasi-fractal $Hom_1(G_S, G_S)$ in the completion by this metric, which is a metric compact with metric d. According to [Danilov V.I., Lectures on fixed points. Russian Economic School, Moscow, 2006—30 p.], this complete metric compactum, when its dimension equals to n, is homeomorphic to the unit ball D_n in R^n, where $D_n = \{x \in R^n | |x| \le 1\}$ and
   $$|x| = |x_1, \ldots, x_n| = \left(x_1^2, \ldots, x_n^2\right)^{\frac{1}{2}}.$$

2. Then, according to Brouwer's theorem, we obtain a fixed point of the compressive map from Example 4, Chap. 2, which defines the predictability level of the system S. In our case, the fixed point of the compressing map is a group of homomorphisms $Hom\left(G_{k+1}^\alpha, G_{k+1}^\alpha\right)$ of a group G_{k+1}^α of some group of the level k of a quasi-fractal $G_S = \langle G_S, *, \square^{-1}\rangle = A_1 = \langle A_1; \Omega_1\rangle$. We shall call $Hom\left(G_{k+1}^\alpha, G_{k+1}^\alpha\right)$ a fixed or invariant semigroup of homomorphisms of a quasi-fractal $Hom_1(G_S, G_S)$.

Remark The fixed semigroup of homomorphisms $Hom\left(G_{k+1}^\alpha, G_{k+1}^\alpha\right)$ of a quasi-fractal $Hom_1(G_S, G_S)$ defines the target functions of the system S.

3.4 Regulating Quasi-fractal of a System. Regulatory Functions of the System. Some Theorems Explaining the Appearance of Mutations

In this section we study the regulation function of the system S from the point of view that the regulation function can be represented as the reaction function of the system S onto the environmental impact on the functioning of the system, determined by its target functions.

3.4.1 Geometric Examples of Quasi-fractal Groups

In this section we consider quasi-fractal systems modeled by quasi-fractal groups and prove the following theorem

Theorem 3.5 *Let the system S be modeled by a quasi-fractal group of permutations of the third degree $S_3 = \langle S_3, *, \square^{-1}\rangle = A_1 = \langle A_1; \Omega_1\rangle$. At any level of the quasi-fractal representing this system it is not possible to regulate exactly:*

- *one factor representing the system,*
- *two factors representing the system,*
- *four factors representing the system.*

When regulating precisely:

- *one factor representing the system,*
- *two factors representing the system,*
- *four factors representing the system,*

system S will change its structure.

So, Theorem 3.5 explains the changes in the structure of the studied system, so one can say that it explains the appearance of mutations.

Theorem 3.6 *Let the system S be modeled by a quasi-fractal the alternating group of permutations of the fourth degree $A_4 = \langle A_4, *, \square^{-1} \rangle = A_1 = \langle A_1; \Omega_1 \rangle$. At no level of the quasi-fractal representing this system it is possible to regulate exactly: the six factors that represent the system. When exactly six factors representing the system are regulated, system S will change its structure.*

3.4.2 The Control Function of the Smart System as a Feedback of the Smart System and the External Environment

In this section we formalize the responses (reactions) of the smart system S to the tests $T = \{T_i | i \in I\}$, using the following definition.

Definition 3.8 Let $f_i : \Gamma \to \Gamma$ be a homomorphism of a graph Γ into itself, which corresponds to a control function. Then the image $f_i(\Gamma)$ is the response of the system $S_\alpha, \alpha \in \Lambda$ to the test T_i, the image $f_\alpha(\{W(T_i) | i \in I\})$ is the response system of the subsystem $S_\alpha, \alpha \in \Lambda$ on the test system $T = \{T_i | i \in I\}$.

3.4.3 Appendix. Quasi-fractal Systems Recognition Methods. Monitoring Testing Outcomes of Smart Model

Chapter 4. Smart Systems' Scales for Measuring

Abstract. In Chap. 4 the following basic questions are considered:

- complex systems' problems of measuring,
- generalizations of classical scales. Multidimensional, matrix and lattice scales, examples.
- coding as a tool to measure students' level of knowledge,
- quasi-fractals and synergistic effects. Scales fixing the synergistic effect and the time (level) series determined by them,
- measurement scales for quasi-fractal algebraic systems.
- quasi-fractal scale measurability level.

Keywords: Testing · Binary tests · System efficiency · Knowledge system · Semantic network · Probability measure

4.1 Introduction

The issues of measuring processes that occur in complex systems modeled by quasi-fractal algebraic systems are discusses in this chapter. The concept of a quasi-fractal scale is defined that allows monitoring and diagnostics in complex systems modeled using quasi-fractal algebraic systems. The level of measurability by a quasi-fractal scale is determined using the Brouwer fixed-point theorem.

4.2 Complex Systems' Problems of Measuring

The problems of mapping and unity of measure can be clarified as follows, [Home page https://en.wikipedia.org/wiki/Level_of_measurement]:

1. The problem of presentation. An empirical system A with relationships is given. Is there a numerical system with relations B, into which one can map homomorphically system A?
2. The problem of uniqueness. Describe the set $\{f\}$ of all homomorphic mappings of the empirical system to the number system.
3. The direct problem of adequacy. Which rules of statistical inference are adequate in a scale with a group of admissible transformations Φ?
4. The inverse problem of adequacy. Statistical inference rule is given. In what scales (that is, under what groups of transformations Φ) is it adequate?

The concept of measurement scales is given in [Pfantsagl I., Theory of measurements, trans. from English, M., 1976; Stevens, S. S. (7 June 1946). 'On the Theory of Scales of Measurement'. Science. 103 (2684): 677–680, Serdyukova NA, Optimization of the tax system of Russia, 2002], in terms of homomorphisms of algebraic systems.

4.2.1 Generalizations of Classical Scales. Multidimensional, Matrix and Lattice Scales

4.2.2 Example. Coding as a Tool to Measure Students' Level of Knowledge

4.3 Quasi-fractals and Synergistic Effects. Scales Fixing the Synergistic Effect and the Time (Level) Series Determined by Them

4.3.1 Explanation of the Choice as an Indicator of the Time (Level) Series Modeling the System of the Number of Synergetic Effects of the Algebraic System of the kth Level of a QUASI-Fractal for Each k

So, one obtains a measurement scale in the form of a fractal time series of synergetic effects of a quasi-fractal system. One can obtain a measurement scale in the form of a fractal time series of synergetic effects of a quasi-fractal system, in another way: to consider the Euclidean metric and mapping inverse to the home-omorphism obtained in the Brouwer fixed- point theorem. Here we use a technic from [Starchenko, N.V., Fractality Index and Logical analysis of chaotic time series, 05.13.18—Mathematical modeling, numerical methods and program complexes 01.01.03—mathematical physics, dissertation for the degree of Ph.D. (physical and mathematical sciences)].

4.4 Measurement Scales for Quasi-fractal Algebraic Systems. Quasi-fractal Scale Measurability Level

The approach based on modern ideas about the formation of a complex system as a dynamic structure that allows you to associate the structural and other qualitative parameters of the system with its quantitative parameters and the form of presentation of information through the fractality category will solve the problems of traditional methods for assessing the properties of the system under study. In [Bavykin, O.B., Fractal multidimensional scale, designed to control the regime of dimensional ECHO and evaluate its output data, Engineering Bulletin, FSBEI HPE MSTU named after N.E. Bauman, 77-48211/596038, No. 07 July 2013], for example, a fractal scale is designed to control the dimensional ECHO mode. To determine the quasi-fractal scale of measurement, we need the definition of a quasi-fractal algebraic model.

Definition 4.4 A quasi-fractal model is a quasi-fractal algebraic system whose signature at all levels of a fractal consists of predicates.

We shall need the Definition 3.1 of the quasi-fractal homomorphism introduced in Chap. 3, and Definition 3.2 of the quasi-fractal semigroup of homomorphisms $\textbf{Hom}_1(\textbf{G}_S, \textbf{G}_S) = A_1 = \langle A_1; \Omega_1 \rangle$ of the quasi-fractal group $\textbf{G}_S = \langle \textbf{G}_S, *, \square^{-1} \rangle$ of factors that determine the system S which is modelling by the group $\textbf{G}_S = \langle \textbf{G}_S, *, \square^{-1} \rangle$. Semigroup operation in $\textbf{Hom}_1(\textbf{G}_S, \textbf{G}_S) = A_1 = \langle A_1; \Omega_1 \rangle$ is composition of homomorphisms. Let's remind the definition.

Definition 4.5 Let $A = \langle A, \Omega \rangle = A_1 = \langle A_1; \Omega_1 \rangle$ be a quasi-fractal model and $B = \langle B, \Omega' \rangle = A_1 = \langle A_1; \Omega_1 \rangle$, where $B \subseteq R$, be a quasi-fractal numerical model. We remind that in this notation $A_1 = \langle A_1; \Omega_1 \rangle$ plays a role of a variable to denote a quasi-fractal algebraic system. An ordered triple $\langle A, B, f \rangle = C_1 = \langle C_1; \Omega_1 \rangle$, in which $\langle f = f_1, f_2 \rangle$ is a quasi-homomorphism from $A = \langle A, \Omega \rangle = A_1 = \langle A_1; \Omega_1 \rangle$ into $B = \langle B, \Omega' \rangle = A_1 = \langle A_1; \Omega_1 \rangle$, is called a quasi-fractal scale.

From Definition 4.5 it follows that a quasi-fractal scale can be constructed in such a way that it would contain all the scales corresponding to the classification of scales and their generalizations—multidimensional, matrix, and lattice scales, (Table 4.1, Chap. 4, this book).

Definition 4.6 A homomorphism $\varphi : R = \langle R, \Omega', \leq \rangle = A_1 = \langle A_1; \Omega_1 \rangle \to R = \langle R, \Omega', \leq \rangle = A_1 = \langle A_1; \Omega_1 \rangle$, where $\leq$ is the relation "less or equal" on the

quasi-fractal R, (that is on the each level of a quasi-fractal) is called an admissible scale transformation of a scale $\langle A, B, f \rangle$, if for an arbitrary fixed homomorphism
$f_0 : A = \langle A, \Omega \rangle = A_1 = \langle A_1; \Omega_1 \rangle \rightarrow B = \langle B, \Omega' \rangle = A_1 = \langle A_1; \Omega_1 \rangle$ the following diagram

$$
\begin{array}{ccccc}
A =< A, \Omega >= A_1 = \langle A_1; \Omega_1 \rangle & \overset{f_0}{\rightarrow} & B = \langle B, \Omega' \rangle = A_1 = \langle A_1; \Omega_1 \rangle & \overset{i}{\rightarrow} & \langle R, \Omega', \leq \rangle = \\
A_1 = \langle A_1; \Omega_1 \rangle & & & & \\
& f \searrow & & \nearrow \varphi & \\
& & B = \langle B, \Omega' \rangle = A_1 = \langle A_1; \Omega_1 \rangle & \overset{i}{\rightarrow} & \langle R, \Omega', \leq \rangle = A_1 = \langle A_1; \Omega_1 \rangle
\end{array}
$$

is commutative, that is $\varphi i f_0 = if$:

$$
\begin{array}{ccccc}
A =< A, \Omega >= A_1 = \langle A_1; \Omega_1 \rangle & \overset{f_0}{\rightarrow} & B = \langle B, \Omega' \rangle = A_1 = \langle A_1; \Omega_1 \rangle & \overset{i}{\rightarrow} & \langle R, \Omega', \leq \rangle = \\
A_1 = \langle A_1; \Omega_1 \rangle & & & & \\
& f \nearrow \quad \varphi i f_0 = if & & \varphi \nearrow & \\
& & B = \langle B, \Omega' \rangle = A_1 = \langle A_1; \Omega_1 \rangle & \overset{i}{\rightarrow} & \langle R, \Omega', \leq \rangle = A_1 = \langle A_1; \Omega_1 \rangle
\end{array}
$$

where $i = \langle i_1, i_2 \rangle$ is a natural embedding of $\langle B, \Omega' \rangle$ into $\langle R, \Omega', \leq \rangle$, that is $i_1 : B \rightarrow R$ is a natural embedding and $i_2 : \Omega' \rightarrow \Omega' \cup \{ \leq \}$ is a natural embedding (that is the diagram is a commutative one at the each level of all quasi-fractals involved in it).

The set of all admissible transformations of a quasi-fractal scale $\langle A, B, f \rangle = C_1 = \langle C_1; \Omega_1 \rangle$ forms a semigroup $\langle \Phi, \circ \rangle$, where $\circ$ is composition operation. An admissible transformation $\varphi \in \Phi$ of a quasi-fractal scale $\langle A, B, f \rangle$ transfer quasi-fractal scale $\langle A, B, f \rangle$ into quasi-fractal scale $\langle A, B, \varphi^\circ f \rangle$ equivalent to it.

Theorem 4.7 *For each quasi-fractal scale, there is a level of measurability of a complex system on this scale.*

4.4.1 Examples of Predicting Numerical Characteristics of Processes Occurring in Complex System

Example. Oil Price Forecasting Methods Here we consider examples related to predicting the numerical characteristics of processes occurring in complex systems and illustrate Theorem 4.7 with these examples.

Chapter 5. Testing Problems. Testing as a Coding of Knowledge System

Abstract. In Chap. 5 the following basic questions are considered: The test system is a measurement system through assessments of students' knowledge. The main issues here are:

- the question of assessing the adequacy of the results of measuring the real level of knowledge and skills of students,
  the question of a comprehensive assessment of the level of assimilation of the system of knowledge that connects quantitative and qualitative indicators.

In this regard we shall consider two models of the algebraic formalization of the representation of the knowledge system—in the form of a free group of factors that determine the knowledge system, and in the form of a semantic network, which we represent using finite Boolean algebra. These models allow us to show that all binary tests i.e. tests involving answers in the form of either "yes" or "no", can be solved true, not knowing the specific nature of the proposed questions. Then we shall show that probability measure can be used as a measure of the level of assimilation of the knowledge system represented by the semantic network

Keywords: Testing · Binary tests · System efficiency · Knowledge system · Semantic network · Probability measure

5.1 Introduction. Main Measurement Problems in Test Field

Many researches claim that tests are a measure of the level of mastery of a knowledge system. However, there are no adequate enough quantitative assessments of the level of mastery of the knowledge system. Under rather strict restrictions, we shall construct such an estimate. It is shown that the tensor estimation of the system efficiency defined in [Serdyukova, N. A., Serdyukov, V. I., Algebraic formalization of smart systems. Theory and Practice, 2018, Chap. 6], including over a field of two elements, can be considered as encoding current state of a system. We shall construct a tensor estimate of the effectiveness of the functioning of the system as a homomorphism of a group of factors , determining the system into a group $GL(n, R)$ of linear homogeneous transformations of the vector space R^n.

The main measurement problems in the field of testing, see also [Program for the Development and Improvement of State Educational Standards and Testing (First Stage) Final Report. Appendix 4. Foreign construction experience and current problems of the development of educational testing], are listed.

5.2 Testing as a Coding of a Knowledge System

Algebraic formalization of smart systems, and, in particular, system identification matrices (in our case, knowledge systems and test systems) allow one to consider testing as a coding of a knowledge system [Solovieva FI, Introduction to Coding Theory, Novosibirsk State University, Novosibirsk, 2006].

5.3 Coding as a Measure Students' Knowledge Tool

Here we proved the main theorem.

Theorem 5.1 *There exists an algorithm that allows one to solve all binary tests of the knowledge system S not knowing the specific essence of the proposed questions.*

5.4 Binary Tensor Estimation of System Efficiency as Encoding the Current State of the System

Here we shall show that a tensor estimate of the effectiveness of a system can be considered as encoding the system's current state. Also, we shall consider a tensor estimate of the effectiveness of a system constructed over a field of two elements generalize this concept as follows.

Definition 5.2 A tensor estimate of the effectiveness of a system S over a field F is a mapping of the group of factors G_S, that determine the system S, into a complete linear group $GL(n, F)$ of the order n, that is the group of all invertible matrices of order n, or the group of invertible linear operators of the space F^n.

Definition 5.3 A binary tensor estimate of the effectiveness of a system S is its tensor estimate over a field of two elements Z_2, that is a mapping of the group of factors G_S, that determine the system S, into a complete linear group $GL(n, Z_2)$ of the order n, that is the group of all invertible matrices of the order n, or the group of invertible linear operators of the space Z_2^n.

So, one obtains from Definition 5.3 that the binary tensor estimate of the effectiveness of the system S is the encoding of the current state of the system S.

We represent the tensor estimate as a homomorphism $G_{S_i} \rightarrow GL(n, R)$, where G_{S_i} is a group of factors defining the system S_i, $GL(n, R)$ is a group of linear homogeneous transformations of a vector space R^n, n—the number of quantitative indicators that assess the quality of the subsystem G_{S_i} of the system S.

5.5 Representation of the Semantic Model of the Knowledge System (Semantic Network of the Knowledge System) in the Form of a Finite Boolean Algebra

Definition 5.4 An algebra $B = \langle\langle F_1, \ldots, F_n\rangle, \wedge, \vee, \rceil, 0, 1\rangle$, where $\wedge$ is a conjunction, $\vee$ is a disjunction, $\rceil$ is a negation, 0 is "false", 1 is "true", is called the closure of the semantic network $\{\{F_1, \ldots, F_n\}, \wedge, \vee, \rceil, 0, 1\}$.

5.5.1 Probabilities on Semantic Networks. Probability Measure as a Measure of the Level of Assimilation of the Knowledge System Represented by the Semantic Network

Here we construct the probability measure p as a measure of the level of assimilation of the knowledge system represented by the semantic network $\{\{F_1, \ldots, F_n\}, , \wedge, \vee, \rceil, 0, 1\}$.

5.5.2 Homomorphisms that Define Measures

Theorem 5.7 *There exists a measure μ on the closure of a finite semantic network $B = \langle\langle F_1, \ldots, F_n\rangle, \wedge, \vee, \rceil, 0, 1\rangle$ which is not a homomorphism of the finite semigroup F, $\cup$ into the semigroup $R^+ \cup \{\infty\} \cup \{0\}/Z^+ \cup \{\infty\} \cup \{0\}$.*

The proof follows from the fact that $\mu(A \cup B) \neq \mu(A) + \mu(B)$, if $A \cap B \neq \emptyset$.

Theorem 5.8 *Any homomorphism $\mu : F \to R^+ \cup \{\infty\} \cup \{0\}/Z^+ \cup \{\infty\} \cup \{0\}$ of the closure of a finite semantic network is a measure on the closure of a finite semantic network.*

Remark Normalizing μ, we get a probability measure, $p = \frac{1}{\mu(\Omega)}\mu, \Omega = \{F_1, \ldots, F_n\}$—is the set of all elementary events of Boolean algebra $2^{\{F_1, \ldots, F_n\}}$.

5.5.3 Quasi-fractal Measure as a Quasi-fractal Homomorphism

Definition 5.9 Let $A_1 = \langle A_1; \Omega_1\rangle$ be a quasi-fractal algebraic system. A quasi-fractal measure on a quasi-fractal system $A_1 = \langle A_1; \Omega_1\rangle$ is a quasi-fractal function $\mu : K = \langle K, \cup, \cap, \backslash\rangle = \langle A_1; \Omega_1\rangle \to R^+ = \langle A_1; \Omega_1\rangle$ defined on some quasi-fractal ring of sets $K = \langle K, \cup, \cap, \backslash\rangle = \langle A_1; \Omega_1\rangle$ on X_k, k is a level of a quasi-fractal ring $K = \langle K, \cup, \cap, \backslash\rangle = \langle A_1; \Omega_1\rangle$ and satisfying the additivity condition on each level $k : \left(\forall A, B \in K_\mu\right)(A \cap B = \emptyset \Rightarrow \mu_k(A \cup B) = \mu_k(A) + \mu_k(B))$.

An ordered quasi-fractal triple $\langle X, K, \mu\rangle = \langle A_1; \Omega_1\rangle$, where $\mu : K \to R^+$ is a quasi-fractal measure, is called a measure quasi-fractal space; quasi-fractal sets from K are called μ-measurable. A quasi-fractal space with measure can also be considered as a pair $\langle X, \mu\rangle = \langle A_1; \Omega_1\rangle$, assuming μ_k-measurable sets to be already defined and giving a ring of sets on X_k. Any measure $\mu_k : K_k :\to R^+$ on X has the following properties:

1. $\mu(\emptyset) = 0$.
2. $(\forall A, B \in K_k)(A \subseteq B \Rightarrow \mu_k(A) \leq \mu_k(B))$ (monotony of measure).
3. If K_k an algebra of sets on X_k, then $\mu_k(X_k)$ is the largest value of the measure μ_k.

Theorem 5.10 *There exists a quasi-fractal measure μ on the closure of a quasi-fractal finite semantic network $B = \langle\langle F_1, \ldots, F_n\rangle, \wedge, \vee, \rceil, 0, 1\rangle$ which is not a homomorphism of the finite quasi-fractal semigroup $\langle F, \cup\rangle = \langle A_1; \Omega_1\rangle$ into the quasi-fractal semigroup $R^+ \cup \{\infty\} \cup \{0\} \backslash Z^+ \cup \{\infty\} \cup \{0\} = \langle A_1; \Omega_1\rangle$.*

The proof follows from the fact that for each level k $\mu_k(A \cup B) \neq \mu_k(A) + \mu_k(B)$, if $A \cap B \neq \emptyset$.

Theorem 5.11 *Any quasi-fractal homomorphism $\mu : F = \langle A_1; \Omega_1\rangle \to R^+ \cup \{\infty\} \cup \{0\} \backslash Z^+ \cup \{\infty\} \cup \{0\}$ of the quasi-fractal closure of a quasi-fractal finite semantic network is a measure on the closure of a quasi-fractal finite semantic network.*

Remark Normalizing μ_k on each level of a quasi-fractal, we get a probability measure, $p_k = \frac{1}{\mu(\Omega)}\mu_k$, $\Omega_k = \{F_{k1}, \ldots, F_{kn}\}$—is the set of all elementary events of Boolean algebra $2^{\{F_{k1}, \ldots, F_{kn}\}}$.

Chapter 6. Smart System's Potential

Abstract. In this Chapter, we introduce the concept of parametric algebraic potential in line with the algebraic formalization of the system.

Keywords: Potential · Algebraic formalization of the system

6.1 Introduction

In various fields of knowledge related to the humanities, such as, for example, economics, teaching theory, control theory, analogues of physical concepts are used, and, in particular, the concept of potential. In this Chapter, we introduce the concept of parametric algebraic potential in line with the algebraic formalization of the system.

Let's remind that the potential energy (in our context is a measure of the transition of motion and interaction of systems or chaos) of a system from one form to another is determined by the mutual position of interacting bodies or parts of the same body, that is, of the structure of the system.

Here we shall use the notions of Borel function and Borel measure.

Definition 6.3 By the algebraization of potential energy or the potential measure of the transition of motion and interaction or the potential property of a system, we call the possibilities determined by the (algebraic) structure of the system.

To count the potential of a system one should construct a flat graph of the lattice of subgroups of the group G_S, then construct an analog of a distance between two vertices of the lattice graph of subgroups of the group G_S, and then use Lagrange theorem to pull on the curve on the vertices of the lattice graph of subgroups of the group G_S.

Definition 6.4 By the algebraization of kinetic energy, or the kinetic measure of the transition of motion and interaction, or the kinetic property of a system, we call the strength of the system's bonds, determined by its (algebraic) structure.

6.2 Main Construction

6.2.1 The Concept of System Potential

We need the following definitions from [Serdyukova, N.A., Serdyukov, V. I., Algebraic formalization of Smart Systems, Theory and Practice, Springer, Smart Innovation, Systems and Technologies, Volume 91, 2018].

Definition 6.5 Let S be a system and G_S be a group of factors that determined the system S. The measure $PC(G_S)$ of the system S links strength is the number of possible different synergetic effects of the system S, that is the number of possible different final states of the system S, which are calculated by the model G_S, or, which is the same, the number of pairwise nonisomorphic groups of order $|G_S|$.

Definition 6.6 Let S be a system and G_S be a group of factors that determined the system S. Let $\varnothing \neq M \subseteq G$. The measure $PC(M)$ of the set M links strength is the number of possible different synergetic effects of the system $\langle G\backslash M\rangle$, where $\langle G\backslash M\rangle$ is a subgroup of the group G_S, generated by the set $G\backslash M$.
1. Let's construct a flat graph of the lattice of subgroups of the group G_S. We arrange ordered pairs at the vertices of the graph (x_V, y_V), where x_V is a reciprocal of the number of synergistic effects of a subgroup of a group G_V, corresponding to vertex V, that is $pc(G_V) = \frac{1}{PC(G_V)}$, and y_V is a communication level i.e. maximum strength between G_V and subgroups of the group G_S, incident to the vertex V in the lattice of subgroups of the group G_S, which is defined as follows:

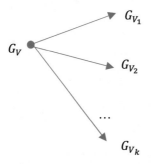

2. We now assume in Definition 6.6 one by one $M = G_V \cup G_{V_i}$, where $i = 1, \ldots, k$, select the maximum value $pc(M) = \frac{1}{PC(M)}$.

$x = pc(G_V)$, $y = pc(M)$, where $M = G_V \cup G_{V_i}$, $i = 1, \ldots, k$. The difference $x - y$—analog of a distance between x и y. It can be done in accordance with the Definition 2.2, Chap. 2.

Now from Definitions 6.5 and 6.6 from [Serdyukova, N.A., Serdyukov, V. I., Algebraic formalization of Smart Systems, Theory and Practice, Springer, Smart Innovation, Systems and Technologies, Volume 91, 2018] we get

Definition 6.7 Using the Lagrange theorem, we pull on the surface (curve) μ in the domain C to the points $\{(x_V, y_V)|V$ runs over the set of vertices ofthe lattice graph of subgroups ofthe group $G_S\}$ and consider the integral.

$$\int_C \frac{d\mu(y)}{|x - y|}$$

The integral $\int_C \frac{d\mu(y)}{|x-y|}$ will be called the algebraization of the potential of the system S according to fhe model G_S.

6.2.2 The Concept of Quasi-fractal System Potential

Now let's introduce the concept of quasi-fractal system algebraic potential as a limit of the algebraic potential of a system of the level k of a quasi-fractal algebraic system at $k \to \infty$.

Definition 6.8 Let $G_S = \langle A_1 = A_1, \Omega_1 \rangle$ be a quasi-fractal group of factors that determine the system S.

Let's construct a flat graph of the lattice of subgroups at each level G_{S_k} with the number $k$of a quasi-fractal $G_S = \langle A_1 = A_1, \Omega_1 \rangle$. We arrange ordered pairs at the vertices of the graph (x_V, y_V), where x_V is a reciprocal of the number of synergistic effects of a subgroup of a group G_{Vk}, corresponding to vertex V, that is $pc(G_{Vk}) = \frac{1}{PC(G_{Vk})}$, and y_{Vk} is a communication level i.e. maximum strength between G_V and subgroups of the group G_{S_k} incident to the vertex Vk in the lattice of subgroups of the group G_{S_k} which is defined as follows:

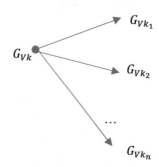

We now assume in Definition 6.6 one by one $M = G_{Vk} \cup G_{Vk_i}$, where
$i = 1, \ldots, n$, select the maximum value $pc(M) = \frac{1}{PC(M)}$.

$x = pc(G_{Vk})$, $y = pc(M)$, where $M = G_{Vk} \cup G_{Vk_i}$, $i = 1, \ldots, n$.

The difference $x - y$—analog of a distance between x and y. It can be done in accordance with the Definition 2.2, Chap. 2.

After that, using the Lagrange theorem, we pull on the surface (curve) μk in the domain Ck to the points $\{(x_{Vk}, y_{Vk}) | Vk$ runs over the set of vertices of the lattice graph of subgroups of the group $G_{S_k}\}$ and consider the integral

$$\int_{Ck} \frac{d\mu(y)}{|x - y|}$$

The limit of integrals $\lim\limits_{k \to \infty} \int_{Ck} \frac{d\mu(y)}{|x-y|}$ will be called the quasi-fractal system S potential according to the quasi-fractal model $G_S = A_1 = \langle A_1, \Omega_1 \rangle$.

6.2.3 Systems Classification by Quasi-fractal System Potential

In accordance with Definition 6.8 one gets the following systems classification by quasi-fractal system potential:

- systems of finite quasi-fractal potential,
- systems of infinite quasi-fractal potential,
- systems without quasi-fractal potential.

Conclusion: Potential can be considered as a measure of smart system reliability. See also [A. I. Podgornaya, R. M. Khusainov, Anti-crisis Potential of fractal Organizations, Bulletin of Kazan Technological University, v.16, (20), 2013, pp. 343–346].

6.3 Physical Meaning of Potential

In this section we try to validate that the physical meaning of the notion of potential is in the following: the potential of a system is a scale to measure the system capacity to change its properties while the system changes its structure.

6.3.1 Potential Theory in Social Sciences

In this section we will examine the experience of using the concept of "potential" in social sciences and in Knowledge Theory.

6.3.2 Measure as a Scale

6.3.3 Probabilistic Measures that Define Measurement Scales

From Chap. 5, Sect. 5.5.2 Homomorphisms that Define Measures, we get the following theorem.

Theorem 6.14 *A finite Boolean algebra with a probability measure is a scale.*
A finite σ-algebra with a probability measure is a scale (defines a scale).

6.4 Examples

Here we consider an example related to economics: a dynamic model of duality of budget—tax system of Russia, using the analogue of the concept of physics that is the notion of system potential—the tax potential of the system.

6.5 Conclusions

Potential is a measure of smart system reliability.

Potential of fractal smart system is a measure of fractal smart system reliability.

Quasi-fractal potential of a quasi-fractal smart system is a measure of a quasi-fractal smart system reliability.

Chapter 7. System Structure Randomness and Integrity. Random Graphs Models

> What is infrastructure? And this is connectedness!
> Raygorodsky. A., Models of random graphs,
> Moscow Publishing House, 2011, p. 43

Abstract. In Chap. 7 the following basic questions are considered:

- random graphs,
- two Erdés–Renyi models,
- analogues of the Erdés–Renyi model for quasi-fractal graphs,
- quasi-fractal homomorphisms, scales, measures (continuation of Chap. 4),
- monitoring the functioning of a system represented by a quasi-fractal random graph,
- system structure randomness and integrity,
- infrastructure, giant component,
- the notion of random quasi-fractal graph (continuation of Chap. 4),
- random quasi-fractal scales (continuation of Chap. 4).

Keywords: Random graphs models · Two Erdés–Renyi models · Infrastructure

7.1 Introduction

The main questions of Chap. 7 will be the randomness of the structure of the system, the infrastructure of the system, its sustainability and the integrity of the system. So, we begin from the difference of the notions "structure" and "infrastructure".

We are interested most of all in the mathematical sense of the words structure and infrastructure.

Definition 7.1 One can define a structure *StrS* of a system *S* as a system of links of the system *S* and an infrastructure *IStrS* of the system *S* as a system of links of the system *StrS*.

Such definition is correlated, for example, with the definition of the infrastructure as code that we have mentioned above.

According to [A. M. Raygorodsky, Models of random graphs and their applications Proceedings of MIPT, 2010. Volume 2, no. 4, pp. 130–140, p. 43] by infrastructure the connectedness of the graph representing the structure of the system under study is understood. We have described in details systems links in [N.A. Serdyukova, V.I. Serdyukov, Algebraic Formalization of Smart Systems. Theory and Practice, 2018, Springer, Chap. 5] using Algebraic Systems Theory and now we would like to concern some aspects of probabilities methods in studying this question.

7.2 Random Graphs

7.2.1 History of the Erdés–Renyi Model of a Random Graph

In this section we concern briefly the history of the Erdés–Renyi model of a random graph.

7.3 Graphs' Probabilistic Properties

Here we continue to consider probabilistic constructions on graphs and algebraic systems. We shall follow [A. M. Raygorodsky, Models of random graphs, Moscow Publishing House, 2011].

7.3.1 Formal Description of the Probability Space Model

We shall use Erdés–Renyi second model [A. M. Raygorodsky, Models of random graphs, Moscow Publishing House, 2011].

Let $V_n = \{v_1, \ldots, v_n\}, n = 1, 2 \ldots$ be a set of vertices. Let's construct a random graph on the set V_n. In addition, we would not consider graphs with multiple edges (multigraphs), graphs with loops (pseudographs), oriented graphs (digraphs). So, it turns out that the possible number of edges for a graph with n vertices is C_n^2. Let's connect any two vertices v_i and v_j with the probability $p_n \in [0; 1]$ independently of all other pairs of vertices. Thus, the edges appear in accordance with the Bernoulli independent tests' scheme tests with a probability of success p_n. So, we get a probability space $\langle \Omega_n, \Sigma_n, p_n \rangle$, in which

$$\Omega_n = \left\{ \Gamma_n = \langle V_n, E_n = \{(u_{v_i}, u_{v_j}) | i, j = 1, \ldots, n\} \rangle \right\}, \Sigma_n = 2^{\Omega_n}$$

is the set of all subsets of the set Ω_n, $p_n(\Gamma_n) = p_n^{E_n}(1 - p_n)^{C_n^2 - |E_n|}$.

7.3.2 The "Almost All Graphs" Property

Let $G(n)$ be the set of all graphs with n vertices, P be some property given on the set of all graphs. We shall use the following designation:

$$G_P(n) = \{\Gamma \in G(n) | P(\Gamma)\} P$$

Definition 7.3 [A.M. Raygorodsky, Models of random graphs, Moscow Publishing House, 2011]

It is said that almost all graphs have the property P, if

$$\lim_{n \to \infty} \frac{|G_P(n)|}{|G(n)|} = 1,$$

and almost no graphs with property$P$, if

$$\lim_{n \to \infty} \frac{|G_P(n)|}{|G(n)|} = 0$$

Definition 7.5 A sub-algebra $\bar{B} = \langle B | \{f_\alpha^{n_\alpha} | \alpha \in \Gamma\} \rangle$ of an algebra $\bar{A} = \langle A | \{f_\alpha^{n_\alpha} | \alpha \in \Gamma\} \rangle$ is called P-pure in $\bar{A}$ (or an embedding φ of a sub-algebra $\bar{B}$ into an algebra $\bar{A}$ is P-pure, if (1) every homomorphism $\bar{B} \xrightarrow{\alpha} \bar{C}$ of the subalgebra $\bar{B}$ into $\bar{C}$ (where $\bar{C}$ is an algebra of the signature $\{f_\alpha^{n_\alpha} | \alpha \in \Gamma\}$ of $\bar{A}$, and (2) $P(\bar{C})$ is true, (3) P is a predicate on the class of algebras of the signature $\{f_\alpha^{n_\alpha} | \alpha \in \Gamma\}$ closed under taking subalgebras and factor algebras, can be continued to a homomorphism β of $\bar{A} = \langle A | \{f_\alpha^{n_\alpha} | \alpha \in \Gamma\} \rangle$ into $\bar{C} = \langle C | \{f_\alpha^{n_\alpha} | \alpha \in \Gamma\} \rangle$ in such a way that the following diagram is commutative:

$$0 \to \bar{B} = \langle B | \{f_\alpha^{n_\alpha} | \alpha \in \Gamma\} \rangle \xrightarrow{\varphi} \bar{A} = \langle A | \{f_\alpha^{n_\alpha} | \alpha \in \Gamma\} \rangle$$

$$\alpha \searrow \qquad \swarrow \beta$$

$$\bar{C} = \langle C | \{f_\alpha^{n_\alpha} | \alpha \in \Gamma\} \rangle$$

(7.1)

that is $\beta\varphi = \alpha$. (A note: The general operations of the same type in algebraic systems will be denoted in identical manner.)

The notion of P—purities of algebraic systems was introduced in [Serdyukova, N. A.: On generalizations of purities, Algebra & Logic, 30, № 4 (1991), pp. 432–456], was studied in [N.A. Serdyukova, V.I. Serdyukov, Algebraic Formalization of Smart Systems. Theory and Practice, 2018, Springer].

Let P be any property given on the set of all algebraic systems $A = \langle A, \Omega \rangle$ of the signature Ω, $G(n, \Omega)$ be a set all algebraic systems of the signature Ω, main set of which consists of n elements, n is natural, $G_P(n, \Omega) = \{A \in G(n) | P(A)\}$.

Definition 7.6 It is said that almost all algebraic systems $A = \langle A, \Omega \rangle$ of the signature Ω have the property P, if

$$\lim_{n \to \infty} \frac{|G_P(n, \Omega)|}{|G(n, \Omega)|} = 1,$$

and almost no algebraic systems $A = \langle A, \Omega \rangle$ of the signature Ω with property P, if

$$\lim_{n \to \infty} \frac{|G_P(n)|}{|G(n)|} = 0$$

Since graphs as algebraic systems can be represented as semigroups the Definition 7.2 is in the frame of Definition 7.5.

Here the problem of finding an analog of the graph connectivity property for algebraic systems arises. We shall clear up a little the analogue of the graph connectivity property for algebraic systems in Chap. 8. Now we note that it concerns operations from the signature of the algebraic system, which become, with some probability, partially defined on the main set of the algebraic system.

7.3.3 Probabilistic Metrics. Connection with Elementary Theories

In this section and the next Sect. 7.3.4 we consider some theorems concentrated in [A.M. Raygorodsky, Models of random graphs, Moscow Publishing House, 2011] which will help one in monitoring the functioning of a system represented by a quasi-fractal random graph.

7.3.4 Sustainability Identification Level of a Quasi-fractal Graph

7.4 Monitoring the Functioning of a System Represented by a Quasi-fractal Random Graph

7.4.1 Description of the Erdés-Renyi Model of a Random Quasi-fractal Finite Graph

In accordance with [A.M. Raygorodsky, Models of random graphs, Moscow Publishing House, 2011] for every level k let's consider the set $V\{_{\alpha_k}^{k}|\alpha_k \in \Lambda_k\} = \{1, \ldots, n_k\}$ of vertices of a random graph. Only edges will be random in the graph.

Let $N_k = C_{n_k}^2$ and $\{e_1, \ldots, e_{n_k}\}$ be edges of a complete graph K_{n_k}. Let $p_k \in [0, 1]$. Let's begin to choose edges from the set $V\{_{\alpha_k}^k | \alpha_k \in \Lambda_k\} = \{1, \ldots, n_k\}$ according to Bernoulli scheme with probability of success p_k, that is, if successful, we include the next edge in the set of edges under construction $E_k = \left\{u_{\alpha_k,\beta_k}^k | \alpha_k, \beta_k \in \Lambda_k\right\}$ of the random graph $\Gamma_k = \left\langle \left\{V_{\alpha_k}^k | \alpha_k \in \Lambda_k\right\}, \left\{u_{\alpha_k,\beta_k}^k | \alpha_k, \beta_k \in \Lambda_k\right\}\right\rangle$, and in case of failure—do not include.

It turns out that, as in the Bernoulli scheme, a graph is a sequence $\omega = (x_1, \ldots, x_{N_k})$ of zeros and units. In fact, the sequence is represented by the graph: $x_i = 1$ means $e_i \in E_k$, if $x_i \neq 1$ then $e_i \notin E_k$. So, we get the probability space $(n_k, p_k) = \left(\Omega_{n_k}, \mathcal{F}_{n_k}, \mathcal{P}_{n_k, p_k}\right)$, in which

$$|\Omega_{n_k}| = 2^{N_k} = 2^{C_{n_k}^2}, \mathcal{P}_{n_k, p_k} = p_k^{|E_k|}(1 - p_k)^{C_{n_k}^2 - |E_k|}$$

Thus, in the Erdös–Renyi model, each edge, independently of all other edges, is included in a random graph of level k with probability p_k. Each event from $\mathcal{F}_{n_k}$ is a set of graphs. So, it is natural to interpret events as properties of graphs. For example: what does it mean to calculate the probability that a graph has a given property? This means that it is required to describe the set A of those graphs that possess the given property, and then find the probability A. If A is a set of all connected graphs, then the probability of the event "the random graph is connected" is equal to $\mathcal{P}_{n_k, p_k}(A) = \sum_{G \in A} \mathcal{P}_{n_k, p_k}(G)$.

Theorem 7.14 *Let the system S is simulated by a finite quasi-fractal algebraic system $A_1 = \langle A_1; \Omega_1 \rangle$ and by a random quasi-fractal finite graph of the first sort $\Gamma_1 = \left\langle \left\{V_{\alpha_1}^1 | \alpha_1 \in \Lambda_1\right\}, \left\{u_{\alpha_1,\beta_1}^1 | \alpha_1, \beta_1 \in \Lambda_1\right\}\right\rangle$, which defines the system S process functioning. Let for any natural k the following equality holds:*

$$p_k = \frac{c \ln k}{k}$$

If $c > 1$, then almost certainly a random graph

$$\Gamma_k = \left\langle \left\{V_{\alpha_k}^k | \alpha_k \in \Lambda_k\right\}, \left\{u_{\alpha_k,\beta_k}^k | \alpha_k, \beta_k \in \Lambda_k\right\}\right\rangle$$

is connected, that is a system S retains integrity property. If $c < 1$, then almost certainly a random graph $\Gamma_k = \left\langle \left\{V_{\alpha_k}^k | \alpha_k \in \Lambda_k\right\}, \left\{u_{\alpha_k,\beta_k}^k | \alpha_k, \beta_k \in \Lambda_k\right\}\right\rangle$ is not connected, that is a system S loses integrity property.

Theorem 7.15 (theorem on the reliability of a system modeled by a finite quasi-fractal algebraic system) *Let a system S is simulated by a quasi-fractal finite algebraic system $A_1 = \langle A_1; \Omega_1 \rangle$ and by a quasi-fractal finite random graph*

$$\Gamma_1 = \left\langle \left\{ V^1_{\alpha_1} | \alpha_1 \in \Lambda_1 \right\}, \left\{ u^1_{\alpha_1,\beta_1} | \alpha_1, \beta_1 \in \Lambda_1 \right\} \right\rangle \text{ of the first sort defining the system } S$$
process functioning. Let for every natural fixed k the following equality holds:

$$p_k = \frac{clnk}{k}$$

If $c \geq 3, k \geq 100$, then
$$\mathcal{P}_{n_k,p_k}\left(\left(\Gamma_k = \left\{ V^k_{\alpha_k} | \alpha_k \in \Lambda_k \right\}, \left\{ u^k_{\alpha_k,\beta_k} | \alpha_k, \beta_k \in \Lambda_k \right\} \text{ is connected}\right)\right) > 1 - \tfrac{1}{k}.$$

7.5 System Structure Randomness and Integrity. Infrastructure and Connectiveness. Quasi-fractal Connectiveness

Definition 7.16 Let system S is simulated by a quasi-fractal finite algebraic system $A_1 = \langle A_1; \Omega_1 \rangle$. Let every level of a quasi-fractal finite graph corresponding to a system $A_1 = \langle A_1; \Omega_1 \rangle$ exists with a probability p. That is at each level k of this quasi-fractal system the graph $\Gamma_k = \left\langle \left\{ V^k_{\alpha_k} | \alpha_k \in \Lambda_k \right\}, \left\{ u^k_{\alpha_k,\beta_k} | \alpha_k, \beta_k \in \Lambda_k \right\} \right\rangle$ with the set of vertices $\left\{ V^k_{\alpha_k} | \alpha_k \in \Lambda_k \right\}$ and with the set of edges $\left\{ u^k_{\alpha_k,\beta_k} | \alpha_k, \beta_k \in \Lambda_k \right\}$ is defined. Here $V^k_{\alpha_k}$ is the vertice of the graph Γ_k, which corresponds to the element $\alpha_k \in \Lambda_k$ of the algebraic system $A^\alpha_k = \langle A^\alpha_k; \Omega_k \rangle, \alpha \in \Lambda_k$, of a fractal level k. We assume that the elements $\alpha_k, \beta_k \in A^\alpha_k$ interact in the system . Then the elements $V^k_{\alpha_k}$ and $V^k_{\beta_k}$ are connected by the edge $u^k_{\alpha_k,\beta_k}$. The quasi-fractal graph $\Gamma_1 = \left\langle \left\{ V^1_{\alpha_1} | \alpha_1 \in \Lambda_1 \right\}, \left\{ u^1_{\alpha_1,\beta_1} | \alpha_1, \beta_1 \in \Lambda_1 \right\} \right\rangle$ which corresponds to this construction is called a quasi-fractal graph of the second sort defining system S functioning.

Definition 7.17 Let system S is simulated by a quasi-fractal finite algebraic system $A_1 = \langle A_1; \Omega_1 \rangle$. Let every level of the quasi-fractal graph defining system $A_1 = \langle A_1; \Omega_1 \rangle$ functioning exists with the probability p. At each level k of this quasi-fractal system the graph $\Gamma_k = \left\langle \left\{ V^k_{\alpha_k} | \alpha_k \in \Lambda_k \right\}, \left\{ u^k_{\alpha_k,\beta_k} | \alpha_k, \beta_k \in \Lambda_k \right\} \right\rangle$ with the set of vertices $\left\{ V^k_{\alpha_k} | \alpha_k \in \Lambda_k \right\}$ and a set of edges $\left\{ u^k_{\alpha_k,\beta_k} | \alpha_k, \beta_k \in \Lambda_k \right\}$ is defined. Here an element $V^k_{\alpha_k}$ is a vertice of the graph Γ_k, which corresponds to an element $\alpha_k \in \Lambda_k$ of the algebraic system $A^\alpha_k = \langle A^\alpha_k; \Omega_k \rangle, \alpha \in \Lambda_k$, of a fractal level k . Let's assume that the elements $\alpha_k, \beta_k \in A^\alpha_k$ interact in the system S with the probability p_k. Then elements $V^k_{\alpha_k}$ and $V^k_{\beta_k}$ are connected by an edge $u^k_{\alpha_k,\beta_k}$ with the probability p_k. The quasi-fractal graph $\Gamma_1 = \left\langle \left\{ V^1_{\alpha_1} | \alpha_1 \in \Lambda_1 \right\}, \left\{ u^1_{\alpha_1,\beta_1} | \alpha_1, \beta_1 \in \Lambda_1 \right\} \right\rangle$ which corresponds to this construction is called a quasi-fractal graph of the third sort defining system S functioning.

From the Erdös–Renyi model we obtain the following theorems.

Theorem 7.18 *Let the system S is simulated by a quasi-fractal finite system* $A_1 = \langle A_1; \Omega_1 \rangle$ *and by a random quasi-fractal finite graph* $\Gamma_1 = \left\langle \left\{ V_{\alpha_1}^1 | \alpha_1 \in \Lambda_1 \right\}, \left\{ u_{\alpha_1, \beta_1}^1 | \alpha_1, \beta_1 \in \Lambda_1 \right\} \right\rangle$, *of the second sort defining system S functioning. Let for every natural fixed k*

$$p = \frac{clnk}{k}$$

If $c > 1$, *then almost certainly a random graph* $\Gamma_k = \left\langle \left\{ V_{\alpha_k}^k | \alpha_k \in \Lambda_k \right\}, \left\{ u_{\alpha_k, \beta_k}^k | \alpha_k, \beta_k \in \Lambda_k \right\} \right\rangle$ *is connected that is the system S retains integrity property. If* $c < 1$, *then almost certainly a random graph* $\Gamma_k = \left\langle \left\{ V_{\alpha_k}^k | \alpha_k \in \Lambda_k \right\}, \left\{ u_{\alpha_k, \beta_k}^k | \alpha_k, \beta_k \in \Lambda_k \right\} \right\rangle$ *is not connected, that is the system S loses fractal property (fractal integrity).*

Theorem 7.19 (theorem on the reliability of a quasi-fractal structure) *Let the system S is simulated by a quasi-fractal finite system* $A_1 = \langle A_1; \Omega_1 \rangle$ *and by a random quasi-fractal finite graph* $\Gamma_1 = \left\langle \left\{ V_{\alpha_1}^1 | \alpha_1 \in \Lambda_1 \right\}, \left\{ u_{\alpha_1, \beta_1}^1 | \alpha_1, \beta_1 \in \Lambda_1 \right\} \right\rangle$, *of the second sort defining system S functioning. Let for every natural fixed k*

$$p = \frac{clnk}{k}$$

If $c \geq 3, n \geq 100$, *then* $\mathcal{P}_{n_k, p_k} \left(\left(\Gamma_k = \left\{ V_{\alpha_k}^k | \alpha_k \in \Lambda_k \right\}, \left\{ u_{\alpha_k, \beta_k}^k | \alpha_k, \beta_k \in \Lambda_k \right\} \text{ is connected} \right) > 1 - \frac{1}{n}.$

Theorem 7. 20 *Let the system S is simulated by a quasi-fractal finite system* $A_1 = \langle A_1; \Omega_1 \rangle$ *and by a random quasi-fractal finite graph* $\Gamma_1 = \left\langle \left\{ V_{\alpha_1}^1 | \alpha_1 \in \Lambda_1 \right\}, \left\{ u_{\alpha_1, \beta_1}^1 | \alpha_1, \beta_1 \in \Lambda_1 \right\} \right\rangle$, *of the third sort defining system S functioning. Let for every natural fixed k*

$$p_k = \frac{clnk}{k}, p = \frac{c'lnk}{k}$$

If $c > 1, c' > 1$, *then almost certainly a random graph*

$$\Gamma_k = \left\langle \left\{ V_{\alpha_k}^k | \alpha_k \in \Lambda_k \right\}, \left\{ u_{\alpha_k, \beta_k}^k | \alpha_k, \beta_k \in \Lambda_k \right\} \right\rangle$$

is connected that is the system S retains integrity property. If $c < 1$, *or* $, c' < 1$ *then almost certainly a random graph*

$$\Gamma_k = \left\langle \left\{ V_{\alpha_k}^k | \alpha_k \in \Lambda_k \right\}, \left\{ u_{\alpha_k, \beta_k}^k | \alpha_k, \beta_k \in \Lambda_k \right\} \right\rangle$$

is not connected, that is the system S loses integrity property.

Theorem 7.21 *Let the system S is simulated by a quasi-fractal finite system* $\mathbf{A}_1 = \langle A_1 ; \Omega_1 \rangle$ *and by a random quasi-fractal finite graph* $\Gamma_1 = \left\langle \left\{ V_{\alpha_1}^1 | \alpha_1 \in \Lambda_1 \right\}, \left\{ u_{\alpha_1,\beta_1}^1 | \alpha_1, \beta_1 \in \Lambda_1 \right\} \right\rangle$, *of the third sort defining system S functioning. Let for every natural fixed k*

$$p_k = \frac{clnk}{k}, \quad p = \frac{c'lnk}{k}$$

$$p_k = \frac{clnk}{k}, \quad p = \frac{c'lnk}{k}$$

If $c \geq 3, n \geq 100$, *then*
$$\mathcal{P}_{n_k, p_k} \left(\left(\Gamma_k = \left\{ V_{\alpha_k}^k | \alpha_k \in \Lambda_k \right\}, \left\{ u_{\alpha_k,\beta_k}^k | \alpha_k, \beta_k \in \Lambda_k \right\} \text{ is connected} \right) \right) > 1 - \frac{1}{n}.$$
If $c' \geq 3, n \geq 100$, *then*
$$\mathcal{P}_{n_k, p_k} \left(\left(\Gamma_k = \left\{ V_{\alpha_k}^k | \alpha_k \in \Lambda_k \right\}, \left\{ u_{\alpha_k,\beta_k}^k | \alpha_k, \beta_k \in \Lambda_k \right\} \text{ is connected} \right) \right) > 1 - \frac{1}{n}.$$

7.5.1 Infrastructure. Giant Component

Example 3 Giant component in knowledge system.

Let's now consider the following example associated with the knowledge system. The integrity of the student's knowledge system maintaining. We can interpret a knowledge system S as a net [Chein, Michel; Mugnier, Marie-Laure (2009). Graph-based Knowledge Representation: Computational Foundations of Conceptual Graphs (http://www.lirmm.fr/gbkrbook/). Springer. https://doi.org/10.1007/978-1-84800-286-9 (https://doi.org/10.1007%2F978-1-84800-286-9). ISBN 978-1-84800-285-2].

Let's begin from the notion of formalization of the concept of "task" in the training system.

Example 4 Formalization of the Concept of "Task" in the Training System.

Let the knowledge system S be represented as the graph. A task is a path in this graph. Recall the definition of a path in a graph. A path in a graph is a sequence of vertices in which each vertex is connected to the next one by an edge.

Definition 7.23 [Home page https://en.wikipedia.org/wiki/Path_(graph_theory), Kuznetsov O.P., Adelson-Velsky G.M., Discrete mathematics for an engineer, M.: Energy, 1980-344 p., Ill. pg. 120–122].

Let Γ be an undirected graph. A path in Γ is such a finite or infinite sequence of edges and vertices $w = (\ldots, a_0, E_0, a_1, E_1, \ldots, E_{n-1}, a_n, \ldots)$, that every two adjacent edges E_{i-1}, E_i have a common vertex a_i.

The set of tasks in the knowledge system S will be denoted as $T = \{T_i | i \in I\}$, the path corresponding to the task $T_i, i \in I$, will be denoted as $W(T_i), i \in I$, and a set of paths in the graph Γ, corresponding to the set of tasks $T = \{T_i | i \in I\}$, will be denoted as $W(T) = \{W(T_i) | i \in I\}$.

One can formalize the knowledge, skills and abilities of a particular student using the notion of graph homomorphism.

Formalization of the concepts of "knowledge", "skills", "skills" of a particular student in the knowledge system S

Let $\{\alpha | \alpha \in \Lambda\}$ be a set of students of the knowledge system S.

Definition 7.24 Let $f_\alpha : \Gamma \to \Gamma$ be a homomorphism of a graph Γ into itself. Then the image $f_\alpha(\Gamma)$ is a the system of knowledge, abilities, skills of the student $\alpha, \alpha \in \Lambda$, the image $f_\alpha(\{W(T_i)|i \in I\})$ is the ability to solve the tasks of the student $\alpha, \alpha \in \Lambda$.

Recall that a homomorphism of a graph Γ is a mapping $f : \Gamma \to \Gamma'$ of vertices of a graph Γ into a set of vertices of a graph Γ' under which the incidence relation is preserved. Let's show now how to formalize attempts to solve the problem by students. To do this, we embed the graph Γ representing the knowledge system S into the complete graph Γ_S. Then we construct the free group G_Γ such that its Van Kampen diagram corresponds to the graph Γ_S.

Let's use the way $W(T_i), i \in I$, in the graph Γ_S and set off corresponding to $W(T_i), i \in I$, part $D(G_\Gamma)$ n the Van Kampen diagram and $H(G_\Gamma)$ in the group G_Γ, then close $H(G_\Gamma)$ up to the subgroup $G(W(T_i))$, which we call the set of attempts to solve the problem T_i. After that we build a homomorphism $f_i : G_\Gamma \to G(W(T_i))$. Then the kernel of this homomorphism $Kerf_i$ is a measure of the deviation of the image of the student's knowledge system from the knowledge system. The random system $f_\alpha(\{W(T_i)|i \in I\})$ should be a connected graph for a student α to be successful.

7.5.2 Infinite Random Graphs

Following [Home page https://en.wikipedia.org/wiki/Rado_graph], here we continue to study Erdős–Rényi model of the second sort.

7.5.3 Target Subsystem of the System. The Hypothesis

The following hypothesis arises: The giant component is the target subsystem of the system.

7.6 Quasi-fractal Random Scales

The following question arises: to what degree will the scale, which has a quasi-fractal construction, preserve its measuring properties during destruction?

7.6.1 Random Quasi-fractal Graph

7.6.2 Random Quasi-fractal Scales

The aim of this section is to construct a notion of a quasi-fractal scale. It is based on the main constructions from Chap. 4.

Examples. Probability space as a scale.

Definition 7.27 Let $A = \langle A, \Omega \rangle = A_1 = \langle A_1; \Omega_1 \rangle$ be a quasi-fractal model and $B = \langle B, \Omega' \rangle = A_1 = \langle A_1; \Omega_1 \rangle$, where $B \subseteq R$, be a numerical quasi-fractal model. Let's remember that in this notation $A_1 = \langle A_1; \Omega_1 \rangle$ plays the role of a variable to denote a quasi-fractal algebraic system. An ordered triple $\langle A, B, f \rangle = C_1 = \langle C_1; \Omega_1 \rangle$, where $f = \langle f_1, f_2 \rangle$ is a quasi-homomorphism from $A = \langle A, \Omega \rangle = A_1 = \langle A_1; \Omega_1 \rangle$ into $B = B, \Omega' == \langle [0, 1], p \rangle = A_1 = \langle A_1; \Omega_1 \rangle$, and p is a probability measure, is called a random quasi-fractal scale.

Definition 7.28 A homomorphfisism $\varphi : R = \langle R, \Omega', \leq \rangle = A_1 = \langle A_1; \Omega_1 \rangle \to R = \langle R, \Omega', \leq \rangle = A_1 = \langle A_1; \Omega_1 \rangle$, where $\leq$ is the relation "equal or less" on the quasi-fractal R, (that is on the each level of the quasi-fractal R) is called an admissible transformation of the random quasi-fractal scale $\langle A, B, f \rangle$, if for every fixed homomorphism $f_0 : A = \langle A, \Omega \rangle = A_1 = \langle A_1; \Omega_1 \rangle \to B = \langle B, \Omega' \rangle = A_1 = \langle A_1; \Omega_1 \rangle$ the next diagram

$$A = < A, \Omega >= A_1 = \langle A_1; \Omega_1 \rangle \xrightarrow{f_0} \langle [0,1], p \rangle = A_1 = \langle A_1; \Omega_1 \rangle \xrightarrow{i} \langle R, \Omega', \leq \rangle =$$
$$A_1 = \langle A_1; \Omega_1 \rangle$$

f φ

$$B = \langle B, \Omega' \rangle = \langle [0,1], p \rangle = A_1 = \langle A_1; \Omega_1 \rangle \xrightarrow{i} \langle R, \Omega', \leq \rangle = A_1 =$$
$$\langle A_1; \Omega_1 \rangle$$

is commutative one, that is $\varphi i f_0 = if$:

$$A = < A, \Omega >= A_1 = \langle A_1; \Omega_1 \rangle \xrightarrow{f_0} \langle [0,1], p \rangle = A_1 = \langle A_1; \Omega_1 \rangle \xrightarrow{i} \langle R, \Omega', \leq \rangle =$$
$$A_1 = \langle A_1; \Omega_1 \rangle$$

f $\varphi i f_0 = if$ φ

$$B = \langle B, \Omega' \rangle = \langle [0,1], p \rangle = A_1 = \langle A_1; \Omega_1 \rangle \xrightarrow{i} \langle R, \Omega', \leq \rangle = A_1 =$$
$$\langle A_1; \Omega_1 \rangle$$

here $i = \langle i_1, i_2 \rangle$ is a natural embedding $\langle B, \Omega' \rangle = \langle [0, 1], p \rangle = A_1 = \langle A_1; \Omega_1 \rangle$ into $\langle R, \Omega', \leq \rangle$, that is $i_1 : B \to R$—is a natural embedding and $i_2 : \Omega' \to \Omega' \cup \{ \leq \}$—is a natural embedding (that is, the commutativity of the diagram takes place at each level of all quasi-fractals involved in it).

It follows from Definition 7.28 that a quasi-fractal scale can be constructed in such a way that it contains all the scales corresponding to Chap. 4, Table 4.1 of the classification of scales and their generalizations—multidimensional, matrix, and lattice scales.

Chapter 8. Cayley Graphs and Probability Isomorphic Groups

"There are also a number of seemingly
unrelated problems from physics,
communication engineering, statistics and so on,
that lead us to consider probabilistic relations
in algebraic structures not equivalent to real line."
Ulf Grenander, Probabilities on Algebraic Structures.

Abstract. In Chap. 8 the following basic questions are considered:

- Cayley diagrams, Cayley complexes and Cayley groups graphs, the relationship between them through the basic concepts of category theory,
- Van Kampen diagrams, relationship of Cayley diagrams and Van Kampen diagrams,
- a brief overview of graph study methods,
- probabilistic properties of graphs,
- the use of first-order logic in the study of graphs,
- category concept, examples, role in mathematics and use in systems theory,
- the relationship between groups and graphs through category theory,
- the possibility of transferring the theory of P-purities of algebraic systems to graphs,
- probability-isomorphic finite quasi-fractal groups as analogues of giant components.

Keywords: Probability-isomorphic groups · Category concept · Giant component

8.1 Introduction

In Chap. 8 we define the concept of probability-isomorphic groups. We introduce this concept using Erdés–Renyi algorithm on finite graphs. To define probability isomorphic groups, we need the concept of a Cayley graph of a group.

In this Chapter, the concept of the Cayley graph of a group and the questions related to this concept are important: Cayley diagrams, Cayley complexes, Van Kampen diagrams, Van Kampen probability diagrams, the connection of Cayley diagrams and Van Kampen diagrams, the connection of the Cayley graph of a group and Van Kampen diagrams through the basic concepts of category theory, category concept, examples, their role in mathematics and their using in systems theory, the relationship between groups and graphs through category theory. The notion of probability-isomorphic groups is introduced. It is an analogue of giant component [Homepage https://en.wikipedia.org/wiki/Giant_component]. We shall consider some notions of Category Theory because many algebraic notions are in deep connection with it: many algebraic objects, such as a group, a monoid, a field of fractions of a domain of integrity, etc. can be represented as a category, [Saunders Mac Lane, Categories for the Working Mathematicians, Second Edition, Springer-Verlag New York, Inc, 1998, 315 pp.]. On the other hand, each category *C* can be considered as a graph, [Saunders Mac Lane, Categories for the Working Mathematicians, Second Edition, Springer-Verlag New York, Inc, 1998, 315 pp]. So, the study of these issues has shown that the problem of identifying a system is extremely important in the study of mathematical objects. Let's begin with significant necessary preliminary notions from [Saunders Mac Lane, Categories for the Working Mathematicians, Second Edition, Springer-Verlag New York, Inc, 1998, 315 pp].

8.1.1 Preliminary Notions. History of the Question

We shall start by examining various approaches to the study of graph theory.

There are many different approaches to study Graph Theory. Let us dwell upon some of them:

- these are first of all combinatorial methods of analyses of graphs—given binary relation on the set of vertices of a graph,
- algebraic methods (especially in enumeration tasks), and foremost methods of algebraic systems, groups theory (especially of substitution groups), semigroups theory, matrix theory methods,
- an elementary theory of category theory is considered in [Saunders Mac Lane, Categories for the Working Mathematicians, Second Edition, Springer-Verlag New York, Inc, 1998, 315 pp, p. 31],
- model's theory methods, [Palyutin EA, Interpretation of graphs in noncommutative theories of Frechet degrees, Fundamental and Applied Mathematics, 2009, Volume 15, No. 2, p. 145–167, 2009, Center for New Information Technologies, Moscow State University, Open Systems Publishing House],
- one can highlight methods of theory of categories, topological and geometric methods, [Saunders Mac Lane, Categories for the Working Mathematicians, Second Edition, Springer-Verlag New York, Inc, 1998, 315 pp, p. 31, Christian Jäkel, christian.jaekel(at)tu-dresden.de Technische Universität Dresden, A Unified Categorical Approach to Graphs, July 2015],
- probabilistic methods are used to study random graphs,
- categories theory methods.

First of all, we want to clear up the connection between Cayley diagrams, Cayley complexes and Cayley graphs of groups, Van Kampen diagrams. Let's begin with connections between Categories Theory and Graph Theory.

8.1.2 Categories Theory Methods in Graph Theory. Approach to Graph Definition through Category Theory

In this section we shall use wonderful book by Saunders Mac Lane [Saunders Mac Lane, Categories for the Working Mathematicians, Second Edition, Springer-Verlag New York, Inc, 1998, 315 pp, p. 31]. Mac Lane used a very clear and accessible description of the concepts of category theory in [Saunders Mac Lane, Categories for the Working Mathematicians, Second Edition, Springer-Verlag New York, Inc, 1998, 315 pp, p. 31], which we shall use.

8.2 Cayley Diagrams, Cayley Complexes and Cayley Groups Graphs

We shall use the definitions of a complex, a skeleton of a graph, a one-dimensional skeleton of a graph in this section and clear up the difference between these notions.

8.2.1 Cayley Diagrams
8.2.2. Cayley Diagram Construction of an Arbitrary Group
8.2.3 Cayley Group Diagram Regard to the Subset of its Elements
8.2.4 Cayley Group Complex
8.2.5 Cayley Planar Complexes
8.2.6 Cayley Group Graph as One-Dimensional Cayley Complex
8.2.7 Cayley Group Graph and Van Kampen Diagram Connection

8.3 Some Other Constructions. *P*—purities in Graphs Theory

Further:

- we generalize the concept of *P*—purities, *P*—pure exact sequences onto graph category, onto quasi-fractal algebraic systems, onto quasi-fractal graphs. This can be done using the fact that a graph is a semigroup. We now single out groups from the semigroup class and consider those graphs that are semigroups that are groups. The theory of*P*—purities in the class of groups can be fully applied to them,
- we can represent a category as a graph, so one can spread *P*—purities theory onto category theory,
- then we embed the graphs in multidimensional Euclidean vector spaces and consider the theory of *P*—pure graph in them.

8.3.1 A Special Case. Multidimensional Graphs in Euclidean Vector Spaces

Definition 8.13 Flat graph $\Gamma_1 = \langle \left\{ V_{\alpha_1}^1 | \alpha_1 \in \Lambda_1 \right\}, \left\{ u_{\alpha_1,\beta_1}^1 | \alpha_1, \beta_1 \in \Lambda_1 \right\} \rangle$, where $\left\{ V_{\alpha_1}^1 | \alpha_1 \in \Lambda_1 \right\}$ is the set of vertices of graph Γ_1, $\left\{ u_{\alpha_1,\beta_1}^1 | \alpha_1, \beta_1 \in \Lambda_1 \right\}$ is the set of edges of graph Γ_1, is called a multidimensional Euclidean flat graph (n - dimensional flat Euclidean graph) if its vertices $\left\{ V_{\alpha_1}^1 | \alpha_1 \in \Lambda_1 \right\}$ are the points of Euclidean vector space that is to every point $V_{\alpha_1}^1, \alpha_1 \in \Lambda_1$, there corresponds an ordered tuple $(x_1^1, \ldots, x_n^1), \alpha_1 \in \Lambda_1$, and the following conditions take place:
(1) to different vertices different n—tuples are corresponded,

(2) there exist $n - 2$—tuple $(x_{i_1}, x_{i+1_1}, \ldots, x_{i_1+n-2})$ such, that for all vertices of graph Γ_1 n—tuples which correspond them have equal values, that is for every $V^1_{\alpha_1}, V^1_{\alpha_2}$ vertices of Γ_1 if

$$V^1_{\alpha_1} \mapsto (x_1, , , x_n), V^1_{\alpha_2} \mapsto (y_1, \ldots, y_n)$$

then $(x_{i_1}, x_{i+1_1}, \ldots, x_{i_1+n-2}) = (y_{i_1}, y_{i+1_1}, \ldots, y_{i_1+n-2})$.

Here $u^1_{\alpha_1, \beta_1} = \left(V^1_{\alpha_1}, V^1_{\beta_1} \right)$ is an edge which connects the vertices $V^1_{\alpha_1}, V^1_{\beta_1}$ in that order.

Now let's normalize all vectors $\left(V^1_{\alpha_1}, V^1_{\beta_1} \right)$ all edges of graphs Γ_1. We get from here that all the concepts of graph theory carry over to the case when the graph Γ_1 is placed in accordance with Definition 8.13 in a Euclidean n—dimensional vector space.

The metric defined in the multidimensional Euclidean vector space $E^n_1 = \langle E^n_1, +, \{\lambda | \lambda \in R\} \rangle$, induces a metric on $\varepsilon \Gamma_1$, where $\varepsilon : \Gamma_1 \to E^n_1$ is the embedding Γ_1 into E^n_1, constructed in Definition 8.13.

Now we embed the quasi-fractal graph $\Gamma_1 = \langle \{ V^1_{\alpha_1} | \alpha_1 \in \Lambda_1 \}, \{ u^1_{\alpha_1, \beta_1} | \alpha_1, \beta_1 \in \Lambda_1 \},$ $Q^1, W^1 \rangle = A_1 = \langle A_1; \Omega_1 \rangle$ which is flat in each its level into multidimensional quasi-fractal Euclidean vector space $E^n_1 = \langle E^n_1, +, \{\lambda | \lambda \in R\} \rangle = A_1 = \langle A_1; \Omega_1 \rangle$ in the following way: i—the coordinate corresponds to the i-th level of the quasi-fractal and for each i, an embedment is carried out according to the Definition scheme 8.13.

The metric defined in the multidimensional quasi-fractal Euclidean vector space $E^n_1 = \langle E^n_1, +, \{\lambda | \lambda \in R\} \rangle = A_{qf} = \langle A_{qf}; \Omega_{qf} \rangle$ induces the metric on $\varepsilon \Gamma_1 = A_1 = \langle A_1; \Omega_1 \rangle$, where $\varepsilon : \Gamma_1 = A_1 = \langle A_1; \Omega_1 \rangle \to E^n_1 = A_{qf} = \langle A_{qf}; \Omega_{qf} \rangle$ is the above embedding $\Gamma_1 = A_1 = \langle A_1; \Omega_1 \rangle$ into $E^n_1 = A_1 = \langle A_1; \Omega_1 \rangle$.

8.3.2 The Relationship between Groups and Graphs from the Outlook of Category Theory

8.3.3 Van Kampen Probability Diagram (Cayley Group Graph). Probability-Isomorphic Groups

We need here to recollect Erdés–Renyi algorithm and the description the Erdés–Renyi random graphs model, see Chap. 7. Following this model, let's embed the Van Kampen diagram $D(G)$ (Cayley graph $\Gamma D(G)$) of a finite, finite quasi-fractal group G transformed into a graph with labels on edges into an n—dimensional vector space R^n. After that let's apply the Erdés–Renyi algorithm and obtain from this the definition of probability-isomorphic groups by binomial distribution.

Definition 8.15 Finite groups (quasi-fractal finite groups) $G = \langle G, *, \square^{-1}, e \rangle$ and $G' = \langle G', \circ, \square^{-1}, e' \rangle$ are called probability-isomorphic ones by binomial distribution if by the model of Erdés–Renyi the probability of the fact that Cayley graph

$\Gamma D(G)$ with $M \subset G, M = G\backslash\{e\}$, of a group G and Cayley graph $\Gamma D(G')$ with $M' \subset G', M' = G'\backslash\{e\}$, of a group G' are equal (isomorphic) is more than 0.5.

Theorem 8.16[1] *Any two finite groups of the same order are probability-isomorphic by regard to the binomial distribution.*

Theorem 8.17 *Any two fractal finite groups of the same order are probability-isomorphic by regard to the binomial distribution.*

Chapter 9. Smart System Sustainability and Smart System Management

> However, in mathematics there are no coincidences.
> Mathematics is the realm of necessity.
> Even the most superficial and casual in appearance
> coincidence always expresses a deeper analogy.
> N. Vavilov, Concrete group theory.
> http://mathscinet.ru/files/VavilovN.pdf

Abstract. Recently the concepts of smart systems, smart education, smart engineering, smart medicine, smart city, smart universities, smart technology and smart communication were introduced in scientific methodology and practice for a modern society. All of these entities are the components of a new system - smart society system. These components must be equipped with corresponding connections that ensure their reliability, sustainability and security, or, in other words, to have an appropriate smart infrastructure. It turns out that we approach the concept of smart infrastructure from the standpoint of tensor estimation. So, the main questions to consider here are the following ones:

- verbal tensor estimates in the Smart Systems Theory,
- methods to increase system reliability,
- tensor estimation of system Q—quality,
- the algorithm of tensor estimation of system Q—quality,
- formalization the approaches and the concept of system sustainability,
- the algorithm of tensor estimation of system sustainability,
- sustainability of a quasi-fractal system (a system modeled by a quasi-fractal algebraic system),
- system sustainability violation: violation of the system's closeness property; malfunction of one of the system's factors,

---

[1]The complexity of the question. It turned out [Cobham, A., Undecidability in group theory, Notices Am. Math. Soc., 9,406, 1962] that the theory of finite groups is not even recursively axiomatizable.

- violation of feedback in the system; factor-flexible quasi-fractal system,
- substitution of system's functions; system's compensational possibilities,
- compensational functions of a factor-flexible quasi-fractal system.

Keywords: System sustainability · Quasi-fractal system · System management

9.1 Introduction. Verbal Tensor Estimates in the Smart Systems Theory

Here we begin with examples.
Example 1
Consider the very phrase SMART[2] This phrase or abbreviation in fact, is about the need to use for tensor estimation in management and project management. Moreover, the phrase SMART, in fact, is a verbal model of the system, written in the short form.
Example 2
Let's consider SWOT analysis, according to [https://en.wikipedia.org/wiki/SWOT_ analysis], SWOT (Strengths: characteristics of the business or project that give it an advantage over others. Weaknesses: relative to others. Opportunities: elements in the environment that the business or project could exploit to its advantage. Threats: elements in the environment that could cause trouble for the business or project analysis (or SWOT matrix) is a strategic planning technique used to help a person or organization identify strengths, weaknesses, opportunities, and threats related to business competition or project planning. This phrase or abbreviation, too, in fact, is about the need to use for tensor estimation in management and project management.
Example 3
Sustainability of complex organizations. For organizations of a complex structure operating in a competitive environment, stability should be understood as the complex property of the system (control object (CO) and its control system (CS)), characterized by the following indicators: survivability—the ability of the system to perform tasks under the deliberate effect of all means of destruction from competitors; stability—the ability of the system to perform tasks in emergency situations; reliability—the ability of the system to perform tasks, maintaining the operability and quality of functioning for a given time (during operation). So, we have abbreviation SSR. This phrase or abbreviation in fact, is about the need to use for tensor estimation for the concept of sustainability. Moreover, the phrase SSR, in fact, is a verbal model of the sustainable system, written in the short form, [Dvoeglazov D.M. The survivability and stability of enterprises of complex

---

[2]The first known use of the term is found in Paul J Meyer in 1965 and later in November 1981 in Management Review by George T. Doran (Doran, G. T. (1981). There's a S.M.A.R.T. way to write management's goals and objectives, [Doran, G. T. (1981). There's a S.M.A.R.T. way to write management's goals and objectives. Management Review, Volume 70, Issue 11(AMA FORUM), pp. 35–36] (Specific, Measurable, Attainable, Relevant, Time-bound).

structure under the influence of external risks, management algorithms and risk adaptation models // Internet journal "SCIENCE" Volume 7, No 1 (2015). http://naukovedenie.ru/PDF/72TVN115.pdf (free access) Zag from the screen. Yaz. Russian, English DOI: 10.15862/72TVN115)].

9.1.1 Methods to Increase System Reliability

Here we give a brief sketch of methods to increase system reliability. We concern the following works, [Serdyukov V.I., Serdyukova N.A., Shishkina S.I. Increase in Products Uptime by using Elements of Artificial Intelligence. Vestn. Mosk. Gos. Tekh. Univ. im. N.E. Baumana, Mashinostr [Herald of the Bauman Moscow State Tech. Univ., Mech. Eng.], 2017, no.1, pp. 62–72. https://doi.org/10.18698/0236-3941-2017-1-62-72, Natalia Serdyukova, Vladimir I. Serdyukov, Sergey S. Neustroev, Svetlana I. Shishkina, Assessing the Reliability of Automated Knowledge Control Results, Conference: 2019 IEEE Global Engineering Education Conference (EDUCON)].

9.2 Tensor Estimation of System Q—quality

Let Q be some quality of a system S, which can be estimated with a help of a set

$$H_Q(S) = \{h_1, h_2, \ldots, h_n\}$$

of numerical characteristics. We shall assume, that these numerical characteristics are functions depending of the time: $h_i = h_i(t), t = 1, \ldots, n$.

We shall construct a tensor estimate of the Q—quality of the functioning of the system S as a homomorphism of a group of factors G_S, determining the system S into a group $GL(n, R)$ of linear homogeneous transformations of the vector space R^n. We shall use the same construction as in [Serdyukova, N.A., Serdyukov, V. I., Algebraic formalization of Smart Systems, Theory and Practice, Springer, Smart Innovation, Systems and Technologies, Volume 91, 2018].

We represent the tensor estimate as a homomorphism $G_S \rightarrow GL(n, R)$, where G_S is a group of factors defining the system, $GL(n, R)$ is a group of linear homogeneous transformations of a vector space R^n, n—a number of quantitative indicators that assess the system Q—quality of the system S.

Let's assume that G_S is a finite group, consisting of n elements: $G_S = \{a_1, a_2, \ldots, a_n\}$ and $h_i = h_i(t) = h_i(a_i, t)$ is a numerical characteristic of a factor a_i.

So, a homomorphism $f : G_S \rightarrow GL(n, R)$ can be considered as the tensor estimation of the quality Q.

Definition 9.1 A homomorphism $f : G_S \rightarrow GL(n, R)$ can be considered as the tensor estimation of the quality Q.

9.2.1 The Algorithm of Tensor Estimation of System Q—quality

9.3 Formalization the Approaches and the Concept of System Sustainability

9.3.1 The Algorithm of Tensor Estimation of System Sustainability

This section is based on the results of [Serdyukova, N.A., Serdyukov, V. I., Algebraic formalization of Smart Systems, Theory and Practice, Springer, Smart Innovation, Systems and Technologies, Volume 91, 2018, Serdyukov V.I., Serdyukova N.A., Shishkina S.I. Increase in Products Uptime by using Elements of Artificial Intelligence. Vestn. Mosk. Gos. Tekh. Univ. im. N.E. Baumana, Mashinostr. [Herald of the Bauman Moscow State Tech. Univ., Mech. Eng.], 2017, no.1, pp. 62–72. https://doi.org/10.18698/0236-3941-2017-1-62-72]. To continue these results, we shall consider sustainability dynamic indices and consider Algorithm 9.2—the algorithm of tensor estimation of system sustainability.

9.4 Sustainability of a Quasi-fractal System (a System Modeled by a Quasi-fractal Algebraic System)

In this section we spread the notion of sustainability to the quasi-fractal systems. We shall consider quasi-fractal systems modeled by quasi- fractal groups.

Definition 9.8 Let $G_S = \langle G_S, \circ, \square^{-1}, e \rangle = A_1 = \langle A_1; \Omega_1 \rangle$ be a quasi-fractal group of factors which represent the system S. Let $G_k^\alpha = \langle G_k^\alpha; \Omega_k \rangle$ be finite for every quasi-fractal level k and $|G_k^\alpha| = n_k$. Let $G_{1S}(k), G_{2S}(k), \ldots, G_{mS}(k)$ be all pair wise non isomorphic groups of n_k elements. Sequence of groups $\langle (G_{1S}(k), G_{2S}(k), \ldots, G_{mS}(k)) | k = 1, \ldots, n, \ldots \rangle$ is called a final state of a system S according the model $G_S = \langle G_S, \circ, \square^{-1}, e \rangle = A_1 = \langle A_1; \Omega_1 \rangle$. A system S is called a final sustainable one according the model $G_S = \langle G_S, \circ, \square^{-1}, e \rangle = A_1 = \langle A_1; \Omega_1 \rangle$ if it has only one final state.

9.5 System Sustainability Violation: Violation of the System's Closeness Property. Malfunction of One of the System's Factors

Theorem 9.9 *Upon termination of the functioning of one of the factors from the group* $G_k^\alpha = \langle G_k^\alpha; \Omega_k \rangle$ *of order* $|G_k^\alpha| = n_k^\alpha$ *representing the level k of the quasi-fractal system S that is equivalent to the removal of this factor from the group of factors* G_k^α, *the system S loses the property of closeness on the level k.*

Theorem 9.10 *Upon termination of the functioning of m factors where m does not divide n_k^α, from the group G_k^α of order $\left|G_k^\alpha\right| = n_k^\alpha$ representing level k of the quasi-fractal system S that is equivalent to the removal of these factors from the group of factors G_k^α, the quasi-fractal system S loses the property of closeness on the level k.*

Theorem 9.11 *A quasi-fractal system S with the group of representing factors G_k^α of the level k of the system S retains the closure property on the level k after the cessation of the functioning factors $\left\{a_i \middle| i \in I_k^\alpha\right\}$ if and only if $\langle G_k^\alpha \backslash \left\{a_i \middle| i \in I_k^\alpha\right\}, \circ, \square^{-1}\rangle$ is a group.*

9.6 Violation of Feedback in the System. Factor-Flexible Quasi-fractal System

Theorem 9.12 *Let the system S be modeled by a quasi-fractal group of permutations of the third degree $S_3 = \langle S_3, *, \square^{-1}\rangle = A_1 = \langle A_1; \Omega_1\rangle$. At any level of the quasi-fractal representing this system it is not possible to regulate exactly:*

– *one factor representing the system,*
– *two factors representing the system,*
– *four factors representing the system.*

When regulating precisely:

– *one factor representing the system,*
– *two factors representing the system,*
– *four factors representing the system,*

system S will change its structure.

Theorem 9.13 *Upon termination of the functioning of one of the factors a or a^{-1} from the group G_k^α representing level k of the quasi-fractal system S, the system S loses the property of feedback on this factor on the level k.*

Theorem 9.14 *Let closed associative system S with a feedback be simulated by a quasi-fractal group of factors isomorphic to alternating group $A_n, n \geq 5$. Upon termination of any of the factors determining the system S, with probability equals to p_1, any subsystem of a system S changes its functioning with probability equals to p_1.*

9.6.1 Substitution of System's Functions. System's Compensational Possibilities

9.6.2 Compensational Functions of a Factor-Flexible Quasi-fractal System

In this section and the next one we shall consider substitution of system's functions and system's compensational possibilities in relation to systems whose models are quasi-fractal algebraic systems. For designations see Chap. 9.

Definition 9.15 Let $\tau = (i, J)$ be a transposition of a degree n. We shall say that the factor a_{ik} replaces the factor a_{jk}, if for the transposition $\tau = (i, J)$ the equality $\varphi(x_1, \ldots, x_n) \equiv \varphi(x_{\tau(1)}, \ldots, x_{\tau(n)})$ is true where $Q_1 \ldots Q_n \varphi(x_1, \ldots, x_n) \in Th(f_{ik})$.

Definition 9.16 Let's say that the system S_{ik} admits the replacements of functions if there exists such transposition $\tau = (i, J)$ of a degree n_k, that $\tau Th(f_{ik}) \neq \emptyset$.

Theorem 9.17 *If for the group of factors $G_{S_{ik}}$ that determine the system S_{ik}, the inequality $\tau Th(f_{ik}) \neq \emptyset$ takes place that is if the group $G_{S_{ik}}$ has a symmetric formula representation then the system S admits the replacement of functions.*

Definition 9.18 Let us say that the system S_{ik} is sustainable by the factors a_{ik} and a_{jk}, if $Th(f_{ik}) = Th(f_{jk})$.

The following theorem is an immediate consequence of the definitions.

Theorem 9.19 *In order for the system S_{ik} to be stable by the factors a_{ik} and a_{jk}, it is necessary that it admits the replacement of functions.*

9.7 Conclusions

The open question here is the question about interchangeability of quasi-fractal system levels during its functioning.

Chapter 10. Interpretation of Solvable and Quasi-fractal Solvable Groups in the Theory of Smart Systems

Abstract. In Chap. 10, we consider smart systems whose models of determining factors are solvable groups. Also, quasi-fractal systems which modeled by quasi-fractal groups of determining factors, the levels of which are solvable groups and simple groups are considered. In Chap. 9, we have considered sustainability issues in relation to systems whose models are quasi-fractal algebraic systems. In this chapter:

- we shall introduce and consider the concept of structural sustainability of a closed associative system with a feedback,
- we shall show that closed innovation systems with commuting determining factors do not have the property of structural stability,
- a closed associative system S with a feedback modeled by a free non-abelian group is structurally sustainable with a minimum regulation range,
- theorems about tensor estimates of structurally sustainable groups.

Keywords: Solvable group · Simple group · Quasi-fractal system

10.1 Introduction. Main Definitions and Finite Groups Classification

Definition 10.1 [Home page https://en.wikipedia.org/w/index.php?title=Special:
ElectronPdf&page=Solvable+group&action=show-download-screen]
   A group G is called solvable if it has a subnormal series whose factor groups (quotient groups) are all abelian, that is
   G is a solvable group if it has a finite set of nested normal subgroups:

$$G \succeq G_1 \succeq G_2 \succeq \ldots \succeq G_n = E$$

such that G_{j-1} is normal in G_j and G_j/G_{j-1} is an abelian group, for $j = 1, 2, \ldots, n$.

10.2 Representation of a Solvable Group as a Quasi-fractal. Risks of Regulation of Systems which Models of Determining Factors are Represented as Solvable Groups

In this section we have represented a solvable group G in a form of a quasi-fractal group.

Theorem 10.2 *If the system S is modeled by a solvable group of factors G_S, then S has a finite number of nested subsystems $H_k, k = 1, \ldots, n$,*
   $H_S = H \succeq H_1 \succeq H_2 \succeq \ldots \succeq H_m = E, m < n$, *such that if with probability $p_i, i = 1, \ldots, m$, the factor $g_i \in G_{i-1} \backslash G_i, i = 1, \ldots, m$, ceases to function, then the subsystems modeled by $H_k, k = 1, \ldots, n$, continue to function without changes. At the same time, G_S continue to function without changes with probability $\prod_{i=1}^{n} (1 - P_i))$. If $p_i, i = 1, \ldots, n$, satisfy the binomial distribution law, then maximum probability $\max(p(g))$ of the failure of the subsystem S achieves for the subsystem of the system S, modeled by $G_{m(g)}$ and*

$$\max(p(g)) = p(g = m(g)) = \frac{n!}{(n - m(g))!m(g)!} p^{m(g)} (1 - p)^{n - m(g)}$$

$$= \begin{cases} \frac{n!}{(n-[(n+1)p])![(n+1)p]!} p^{[(n+1)p]} (1 - p)^{n-[(n+1)p]}, & if\ m(g) = [(n+1)p] \\ \frac{n!}{(n-([(n+1)p]-1))!([(n+1)p]-1)!} p^{([(n+1)p]-1)} (1 - p)^{n-([(n+1)p]-1)}, \\ \quad if\ m(g) = [(n+1)p] - 1 \end{cases}$$

   One may consider $f_{i-1} : G_{i-1} \to G_{i-1}/G_i \to 0$, and then f_{i-1} can be considered as a function of the subsystem G_{i-1} of the system G.

10.2.1 Theorems on Simple Finite Groups' Classification

In this section, we briefly touch on the main theorems concerning the classification of finite groups, such as Classification Theorem, Burnside Theorem, Feith–Thompson Theorem, Hall Theorem, Wielandt Theorem.

10.3 Structurally Sustainable Groups

10.3.1 The Physical Meaning of the Concept of System Structural Sustainability

Theorem 10.3 *A closed associative system, with feedback and commuting factors, i.e., modeled by the abelian group of determining factors G_S, is structurally unsustainable.*

Theorem 10.4 *A closed associative system S with a feedback modeled by a free non-abelian group is structurally sustainable with a minimum regulation range.*

Definition 10.5 The equation

$$w(x_1, \ldots, x_n; g_1, \ldots, g_m) = 1 \qquad (10.1)$$

of n variables with coefficients in the group G is determined by the element w of the free product $G * F(x_1, \ldots, x_n), g_1, \ldots, g_m \in G$.

If G is a subgroup of K, then under the solution of the Eq. (10.1) in K we mean a tuple $(a_1, \ldots, a_n)$ of elements of K such that $w(a_1, \ldots, a_n; g_1, \ldots, g_m) = 1$ in K that is, w goes into unity under the homomorphism $G * F(x_1, \ldots, x_n) \to K$, which maps x_i into a_i and identical on G. If $K = G$, then one speaks about the solutions of the equation (10.1) in G.

Definition 10.6 A group G is called algebraically closed if every finite system of equations compatible over G already has a solution in G. A group G is called strongly algebraically closed if every system of equations is a compatible over G in the case if every its finite subsystem is a compatible over G.

Definition 10.7 A subgroup A of a group G is called a pure subgroup of G, if every system (not only finite systems, as in [A. Yu. Olshansky, A. L. Shmelkin, Infinite groups, Itogi Nauki i Tekhniki. Ser. Lying. prob. mat. Fundam. Directions, 1989, Volume 37, 5–1136. Mazhuga, A. M., Verbally closed subgroups, Thesis for the degree of candidate of physical and mathematical sciences, 2018]) of equations over A, that has a solution in G, has a solution in A too.

Theorem 10.8 *Let $E \to A \xrightarrow{\varepsilon} G$ be a pure embedding of a group A ϭ into a group G, where E is a unit group, and A and G be arbitrary groups. If A is a strongly algebraically closed group, then εA is a retract of G.*

In this section we use a version of the compactness theorem [Keysler, Chen, Model Theory, 1974], related to ultraproducts.

Definition 10.9 Let G be a group. Let's consider all finite systems of equations over G. Let

$w_i(x_1, \ldots, x_n; y, g_1, \ldots, g_m) = 1, i \in I$, be all systems of equations of n variables with coefficients in the group G over G having decision in G.

Theorem 10.10 *A closed associative system S with a feedback modeled by a strongly algebraically closed group is structurally sustainable with a minimum regulation range.*

Definition 10.11 [Gupta, CY. K., Romanovsky, N. S.: Noternity by some equations solvable groups, Algebra and Logic, 46, No. 1 (2007), 46–59].

A group A is Noetherian by equations if every system of equations from $x_1, \ldots, x_n$ with coefficients from A is equivalent to some of its finite subsystem over A for any natural number n.

Theorem 10.12 *Every finitely generated group with a solvable word equality problem is embeddable in any algebraically closed group.*

Here from one can get the following theorem.

Theorem 10.13 *Any closed associative system S with a feedback can be embedded into an infinite structurally sustainable system.*

10.4 Tensor Estimation on Smart Systems Modelling by Finite Simple and Finite Solvable Groups

10.4.1 Tensor Estimates of Structurally Sustainable Groups
Theorem 10.14 (On tensor estimation of a closed associative system with a feedback)

Let S be a closed associative system with a feedback, which is modeled by a group of determining factors G_S. If there exist an epimorphisms $\varphi_3 : F_2 \to G_S$, where F_2 is a free non abelian group of a rank 2, (that is, for example, a group G_S has two generating elements, for example system S has only two determining factors) and an epimorphism $\varphi_4 : GL(n, R) \to G_S$, then tensor estimation

$$h : G_S \to GL(n, R)$$

of the system S (according to a model G_S) coincides with the restriction of tensor estimation of a strongly algebraically closed group by the set G_S:

$$f : SA(G_S) \to GL(n, R)$$

that is $h = f \restriction G_S$. So, if a closed associative system with a feedback is modeled by such a group of determining factors G_S, that there exists epimorphisms of a free non abelian free group of a rank 2 onto G_S and of a $GL(n, R)$ onto G_S, then the tensor estimation of a system S is a restriction of a tensor estimation of its supersystem modeled by a strongly algebraically closed group.

Theorem 10.15 (on tensor estimation of structurally sustainable systems with a minimum regulation range) *Let S be a closed associative system with a feedback, which is modeled by a group of determining factors G_S and $\varphi_1 : F_\omega \to G_S$ be an epimorphism of a free nonabelian group of countable rank F_ω onto G_S. If there exists an epimorphism $\varphi_2 : GL(n, R) \to G_S$ then tensor estimation h: $G_S \to GL(n, R)$ of the system S (according to a model G_S) coincides with the restriction of tensor estimation of a strongly algebraically closed group by the set G_S:*

$$f : SA(G_S) \to GL(n, R)$$

that is $h = f \upharpoonright G_S$. So, if a closed associative system with a feedback is modeled by such a group of determining factors G_S, that there exists epimorphisms of $GL(n, R)$ onto G_S, then the tensor estimation of a system S is a restriction of a tensor estimation of its supersystem modeled by a strongly algebraically closed group.

Chapter 1. Problem statement. The connection between the problem of smart systems identification and classical logic.

Chapter 2. Quasi-fractal algebraic systems as a possible approach to solving the problem of smart systems identification

Chapter 3. General system functions. Target subsystems of smart systems. System predictability levels.

Chapter 4. Smart System Measurement Scales. Quasi-fractals and synergistic effects. Quasi-fractal scale measurability

Chapter 5. Testing as smart system encoding (on the example of a knowledge system). Quasi - fractal measures and quasi-fractal homomorphisms.

Chapter 6. The concept of a parametric algebraic potential of a system. The potential of a quasi-fractal system. Classification of smart systems by potential. Probabilistic measures defining scales.

Chapter 7. System structure. Randomness and integrity. Technique Erdés - Renyi. The target subsystem of the system.

Chapter 8. Cayley graphs and probability-isomorphic groups.

Chapter 9. Verbal tensor estimates in the Smart Systems Theory. Sustainability of a system modeled by a quasi- fractal algebraic system, substitution of system's functions; system's compensational possibilities

Chapter 10. The concept of structural sustainability of a closed associative system with a feedback, structurally sustainable and unsustainable smart systems, structurally sustainable systems with a minimum regulation range.

The connections between the first four chapters represent a complete graph:

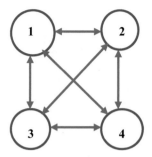

The connections between others chapters run as follows:

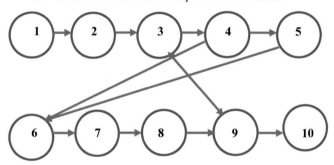

Moscow, Russia   Natalia A. Serdyukova
                                                                  Vladimir I. Serdyukov

Contents

Chapter 1
Smart Systems Identification Problem

Abstract In the first Chapter, the problem of identification of a smart system is considered from the standpoint of compliance of the functioning of the system with its purpose of functioning and evaluation of this compliance. The main issues here are: In the first Chapter, the problem of identification of a smart system is considered from the standpoint of compliance of the functioning of the system with its purpose of functioning and evaluation of this compliance. The problem of identification of smart systems is to answer the question: does the system satisfy the objective of its operation at each moment in time, and if so, how much? Thus, the problem of identification is divided into the following two tasks:

1. Does the system S satisfy the objective of its operation $P(S)$?
2. To assess how the system S satisfies the goals of its functioning $P(S)$?

Also, we construct a structural (by the structure of system attributes) analogue of the econometric method of principal components, developed by the outstanding Russian algebraist D. K. Faddeev.

Keywords Identification problem · Smart system · Identification matrix

1.1 Introduction

Main idea of this chapter runs as follows:

The purpose of the functioning of the economic system is to become a smart system.

The purpose of the functioning of the financial system is to become a smart system.

The purpose of the functioning of the training system is to become a smart system. We distinguish classes of systems whose purpose of functioning is to become a smart system, or come to the final state of smart.

© Springer Nature Switzerland AG 2021
N. A. Serdyukova and V. I. Serdyukov, *Algebraic Identification of Smart Systems*, Intelligent Systems Reference Library 191,
https://doi.org/10.1007/978-3-030-54470-6_1

1.2   The Problem Statement. Problems of Identification and Pattern Recognition: What is the Difference

Pattern recognition is the task of classification and consists in sorting an object to a particular class by isolating the essential features that characterize these data from the total mass of irrelevant data, [1, 2].

The problem of identification of smart systems is to answer the question: does the system satisfy the objective of its operation at each moment in time, and if so, how much? Thus, the problem of identification is divided into the following two tasks:

1. Does the system S satisfy the objective of its operation $P(S)$?
2. To assess how the system S satisfies the goals of its functioning $P(S)$?

The second question comes determining the measure of conformity of the functioning of the system to its goal. Note that in [3–7] a partial solution of the second question based on the use of probability theory methods was proposed.

Let's now explain the wording of the identification problem by examples from [8].

Example 1.1 The problem of human identification. The human genome solves the problem of identification in this case.

Example 1.2 The functioning of the knowledge system. Let's consider the following scheme (Fig. 1.1).

The reason for the latter is the inability not only to solve, but also to pose the problem of identification. One of the aspects of modeling is the determination of characteristics for the model of the system under study. Establishing the characteristics of a model is called identification.

The very concept of identification comes from the Latin word identify—identify. To identify an object means to recognize it, that is find out which class it belongs to. The theory of pattern recognition and its sections—the theory of classification, the theory of signs—investigate identification algorithms. Therefore, the theory of identification is closely connected with the theory of algorithms.

One of the purposes of identification, one can say, the simplest task of identification is to establish the correspondence of a recognized object to its image or a sign which is called an identifier. The problem of structural identification is associated

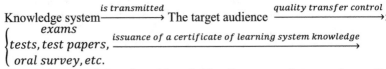

Completion of training →working → The discrepancy between the quality of the work results and the requirements for the results.

Fig. 1.1 A scheme for Example 1.2

with the reflection of the structural properties of the system under study. Aspects of the identification problem are the definition of the composition of the system elements, their classification, unification and typing of standard or typical sets of elements. The identification of the structure requires the solution of the problem of the system's decomposition or its individual subsystems' decomposition, synthesizing parts of the system in accordance with the goal of creating a model, generalizing and analyzing information about the relationships and interactions of elements of the system and interactions of its subsystems, or of interactions of different levels of the system which is under study. The complexity of the problem of identification of the system is that in solving these problems, some details of the studied objects are excluded from consideration. Therefore, one needs to have data on the relative importance of the individual components of the system and their relationships. The identification process has a close relationship with informative, informal, heuristic conclusions. This is especially evident in the first stages of the structural analysis of complex systems, carried out in conditions of uncertainty—the conditions of insufficient information compared with the specified purpose of the analysis.

The essence of identification is to determine the organization of the model, its functional properties or behavior in specific conditions.

Let's try to formalize the statement of the problem.

1.2.1 Identification Problem Formalization

Let the system S consists of elements $\{a_k | k \in K_S\}$, subsystems $\{S_i | i \in I_S\}$, links of different levels $\{l_j | j \in J_S\}$.

In the process of the system S functioning its model S' appears, that is our image of the system S. Let the system S' consists of elements $\{a'_k | k \in K_{S'}\}$, subsystems $\{S'_i | i \in I_{S'}\}$, links of different levels $\{l'_j | j \in J_{S'}\}$.

The systems S and S' differs from each other. Differences can be at the level of:

(1) elements,
(2) subsystems,
(3) links.

Suppose that at the level of the elements $\{a_\gamma | \gamma \in \Gamma_S\} \subseteq \{a_k | k \in K_S\}$, and no one of the elements a_γ, $\gamma \in \Gamma_S$, belongs to the set $\{a'_k | k \in K_{S'}\}$. Suppose that at the level of the subsystems: $\{S_i | i \in \Lambda_S\} \subseteq \{S_i | i \in I_S\}$ and no one of the S_i, $i \in \Lambda_S$, belongs to the set $\{S' | i \in I_{S'}\}$. Suppose that at the level of links $\{l_j | j \in B_S\} \subseteq \{l_j | j \in J_S\}$, and no one of the links l_j, $j \in B_S$, belongs to the set $\{l'_j || j \in J_{S'}\}$.

At a certain point in time, systems S and S' begin to function. How one can understand that systems S and S' have the same properties or there is a difference between the properties of the system S and the properties of the system S'? How can one can estimate the range of these differences?

1.2.2 Possible Approaches to Solve Identification Problem. System Identification Up to Internal or External Attributes

The idea of solving the problem defined by the title of this section is to determine the equivalence relation on the set of predicates and then use the analogy of the introduced concepts with the notion of quasi-isomorphism of Abelian groups, [3, 9].

Let $P(S)$ be the integrity property of the system S, $\{Q_i | i \in I_S\}$ be the external attributes of the system S, that is $\vDash P \rightarrow \wedge_{|i \in I_S} Q_i$, $\{Q_j | j \in J_S\}$ be the internal attributes of the system S, that is $\vDash \wedge_{|i \in J_S} Q_j \rightarrow P$, [3]. Here $Ou_S = \{Q_j | j \in J_S\}$ is the set of unary predicates on the set of all algebraic systems $\Re$ of the same signature Ω(or on the set of all groups), defining the system internal attributes of the system S, $In_S = \{Q_i | i \in I_S\}$ is the set of all unary predicates on the set of all algebraic systems of the same signature Ω (or on the set of all groups defining the system internal attributes of the system S. That is:

$$Q_i: \Re \rightarrow \{0, 1\}, i \in I; \; Q_j: \Re \rightarrow \{0, 1\}, j \in J$$

Let $M = \{Q_i | i \in I\} \cup \{Q_j | j \in J\}$

Let's define a group structure on M (or consider all possible groups on M):

Let $\mathcal{M} = \langle M, *, \square^{-1}, E \rangle$ be a group, E be a predicate acting as a unit of the group $\mathcal{M}$, that is.

$$Q_k * E = E * Q_k = Q_k, k \in I \cup J, \; Q_k * Q_k^{-1} = E = Q_k^{-1} * Q_k$$

The following cases are possible:

(1) The set M of all external and internal attribute attributes of the system S is finite. In this case, we consider all groups up to isomorphism on M. The following cases are possible:

1. The number of elements of the group $\mathcal{M} = \langle M, *, \square^{-1}, E \rangle$ is finite and simple.
2. The group $\mathcal{M} = \langle M, *, \square^{-1}, E \rangle$ is a simple one of the order less than 10^6. In each of these cases, a unique up to isomorphism group $\mathcal{M} = \langle M, *, \square^{-1}, E \rangle$ is obtained.

(2) The set M of all external and internal attributes of the system S is infinite.

Now let $\{S_\gamma | \gamma \in \Gamma\}$ be a set of all subsystems of the system S. Let's suppose that $\vDash (Q_k(S_\gamma) \wedge Q_j(S_\gamma)) \Rightarrow \vDash (Q_k * Q_j)(S_\gamma)$, where $\wedge$ is a conjunction of predicates.

Let $f: \mathbb{M} = < M, \Lambda > \rightarrow \mathcal{M}' = < M, * > $—be a homomorphism of associative semigroups such that $\vDash (Q_k(S_\gamma) \wedge Q_j(S_\gamma)) \Rightarrow \vDash (f(Q_k)) * (f(Q_j))(S_\gamma)$.

Elements of the kernel $Kerf$ not distinguished. The system is identified by a set of properties contained in the kernel $Kerf$: when one disables any property from $Kerf$ the system loses the entire its unique set of properties from $Kerf$.

In order to connect the external and internal system features with statistical calcu-lations, we shall construct probability spaces on predicate algebras that define the external and internal features of the system.

Let again

$In_S = \{Q_i | i \in I_S\}$ be a set of unary predicates on the set of all algebraic system of the same signature Ω (or on the set of all groups), defining system internal attributes S. That is:

$$Q_i: \Re \to \{0, 1\}, i \in I; \quad Q_j: \Re \to \{0, 1\}, j \in J$$

Let $M = \{Q_i | i \in I\} \cup \{Q_j | j \in J\}$, $B(In_S)$ be the closure of the set In_S with respect to the operations of taking all countable conjunctions of predicates from the set In_S, and of taking all finite disjunctions of predicates from the set In_S. Then the algebra.

$\mathcal{M}_1 = \langle B(In_S), \vee, \wedge \rangle$ is a σ-algebra. Let's define on the σ—algebra $\mathcal{M}_1 = \langle B(In_S), \vee, \wedge \rangle$ a probability measure p, that is the function $p_1: B(In_S) \to [0; 1]$, which satisfies the properties:

$p_2(\Omega) = 1$, where Ω is a predicate which takes values on the set of all algebraic system of the same signature Ω (or on the set of all groups), one of which is an algebraic system of factors determining the system S;
$p_2(\bigvee_{i=1}^{\infty} Q_i) = \sum_{i=1}^{\infty} p_1(Q_i)$.

Definition 1.1 A σ—algebra $\mathcal{M}_1 = \langle B(In_S), \vee, \wedge \rangle$ is called the probabilistic space of internal attributes of the system S.

Let $M = \{Q_i | i \in I\} \cup \{Q_j | j \in J\}$, $B(Ou_S)$ is the closure of the set $Ou_S = \{Q_j | j \in J\}$ of all external attributes of the system S with respect to the oper-ations of taking all countable conjunctions of predicates from the set Ou_S, and of taking all finite disjunctions of predicates from the set Ou_S. Then the algebra $\mathcal{M}_2 = \langle B(Ou_S), \vee, \wedge \rangle$ is a σ—algebra. Let's define on the σ—algebra $\mathcal{M}_2 = \langle B(Ou_S), \vee, \wedge \rangle$ a probability measure p, that is the function $p_2: B(Ou_S) \to [0; 1]$, which satisfies the properties:

1.1 $p_1(\Omega) = 1$, where Ω is a predicate which takes values on the set of all algebraic system of the same signature Ω (or on the set of all groups), one of which is an algebraic system of factors determining the system S;

1.2 $p_1\left(\vee_{j=1}^{\infty} Q_j\right) = \sum_{j=1}^{\infty} p_2(Q_j)$

Definition 1.2 A σ-algebra $\mathcal{M}_2 = \langle B(Ou_S), \vee, \wedge \rangle$ is called the probabilistic space of external attributes of the system S.

1.3   System Description up to External Attributes. (Systems Classification up to External Attributes)

1. **Method.** Let's construct an algebra $\langle Ou_S | \wedge, E \rangle$, where $\wedge$ is a conjunction of predicates E is a unary predicate identically true on $\Re$. Then $\langle Ou_S | \wedge, E \rangle$ is an associative semigroup with a unit. The set of all congruences on the algebra $\langle Ou_S | \wedge, E \rangle$ defines the classification of the system up to its external attributes. Let's construct an algebra $\langle In_S | \wedge, E \rangle$, where $\wedge$ is a conjunction of predicates E is a unary predicate identically true on $\Re$. Then $\langle In_S | \wedge, E \rangle$ is an associative semigroup with a unit. The set of all congruences on the algebra $\langle In_S | \wedge, E \rangle$ defines the classification of the system up to its internal attributes.
2. **Method.** Let's define the structure of an algebraic system of signature Ω on the set Ou_S, or accordingly, the structure of the group in all possible ways. Let, for example, $\langle Ou_S |*, \square^{-1}, e \rangle$ be a group obtained in such a way.

All normal subgroups of the group $\langle Ou_S |*, \square^{-1}, e \rangle$ define all possible congruences on $\langle Ou_S |*, \square^{-1}, e \rangle$. Thus, we obtain a description of the system up to its external attributes.

Let's define the structure of an algebraic system of signature Ω on the set In_S, or accordingly, the structure of the group in all possible ways. Let, for example, $\langle In_S |*, ^{-1}, e \rangle$ be a group obtained in such a way.

All normal subgroups of the group $\langle In_S |*, \square^{-1}, e \rangle$ define all possible congruences on $\langle In_S |*, \square^{-1}, e \rangle$. Thus, we obtain a description of the system up to its internal attributes.

As an application of this technics, one can single out a finite simple number of attributes and construct the corresponding models (groups of factors G_S, that determine the system S), that clearly identify the system, since there are no submodels (in the sense of algebraic subsystems) in this case.

Examples

1. The number of elements of the group $\langle \boldsymbol{Ou}_S |*, \square^{-1}, \boldsymbol{e} \rangle$ is finite and simple.
2. A group $\langle \boldsymbol{Ou}_S |*, \square^{-1}, \boldsymbol{e} \rangle$ is a simple one of the order less then 10^6.
3. Test system for knowledge control.

As to concern to the third example, the following questions arise:
How is this system described in terms of systems theory axiomatics?
How one can define the knowledge system integrity property?
How one can determine the external attributes of a knowledge system?
How one can determine the internal attributes of a knowledge system?

1.3.1 Examples. Test System for Knowledge Control. Connection of System Identification Problem with Classical Logic

The attributes of the knowledge system include the breadth of knowledge, the depth of knowledge, the validity of tests.

Example 1. Test System for Knowledge Control Let's consider the equation

$$a \cdot x = b$$

Now let's give the correct algorithm for solving this equation:

Algorithm 1.1

$$a^{-1} \cdot (a \cdot x) = a^{-1} \cdot b$$
$$\left(a^{-1} \cdot a\right) \cdot x = a^{-1} \cdot b$$
$$e \cdot x = a^{-1} \cdot b$$
$$x = a^{-1} \cdot b$$

In accordance with this algorithm, the knowledge base consists of the operation of multiplication $\cdot$ and its properties, the operation of taking the inverse element a^{-1} and its properties, the selected element e the nullary operation and its properties, as well as the rules for sequentially performing these operations or the composition operation.

Let's define the following predicates.

Let predicate E highlights all the rules related to the unary operation of selecting a neutral element, for example, $E: a^{-1} \cdot a = e$ (the predicate is the neutral element). Let predicate A_1 selects all the rules related to the operation of multiplication, for example, $A_1: a^{-1} \cdot (a \cdot x) = a^{-1} \cdot b$ (the rule of multiplying both sides of the equality by the same element a^{-1}). For the predicate A_1 we also have:

$A_1: a^{-1} \cdot (a \cdot x) = \left(a^{-1} \cdot a\right) \cdot x$ (associativity of the operation of multiplication).

$A_1^{-1}: a^{-1} \cdot (a \cdot x) = \left(a^{-1} \cdot a\right) \cdot x$ (predicate of associativity of the operation of multiplication).

Let $A_1^{-1} = A_1$, $E^{-1} = E$.

For example,

$A_1^{-1}: a \cdot \left(a^{-1} \cdot (a \cdot x)\right) = a\left(a^{-1} \cdot b\right)$ multiplication by $a = (a^{-1})^{-1}$ (associativity of the operation of multiplication).

So, we get a cyclic group of the second order Z_2 with the Cayley table (Fig. 1.2):

Fig. 1.2 Cayley table for a cyclic group of the second order Z_2 for Algorithm 1

$\circ$	E	A_1
E	E	A_1
A_1	A_1	E

Fig. 1.3 Cayley table for a
cyclic group of the second
order Z_2 for Algorithm 2

∘	O	A_2
O	O	A_2
A_2	A_2	O

A group Z_2 describes Algorithm 1.

Let's now consider the equation $a \cdot x + b = c$.

We give the correct algorithm for solving this equation:

Algorithm 2

$$(a \cdot x + b) + (-b) = c + (-b)$$
$$ax + (b + (-b)) = c + (-b)$$
$$ax + 0 = c + (-b)$$
$$ax = c + (-b)$$

And, taking into account Algorithm 1, we have:

$$x = a^{-1}(c + (-b))$$

Let predicate A_2 highlights all the rules related to the operation of addition, for example, for example, associativity of addition: $A_2: (a + b) + c = a + (b + c)$, predicate O selects all the rules related to the nullary operation of selecting a neutral element with respect to the addition operation. Let

$$A_2^{-1} = A_2, \, O^{-1} = O$$

So, we get a cyclic group of the second order Z_2 with the Cayley table (Fig. 1.3).

Identification of predicates A_1 and A_2, that is, the case when a student mixes up the operations of multiplication and addition and neutral elements with respect to these operations, leads to errors in solving the equation. A group $Z_2 \times Z_2$ describes Algorithm 2.

Example 2. Connection of System Identification Problem with Classical Logic
Let's consider the classic logic and the set of statements of the form $A \rightarrow B$, where A means the statement "Infections exist," the statement B means the statement "Everyone will die as a result of the infection." This statement is false from the point of view of classical logic, because someone can die in an accident. But if we identify with the help of a unary predicate given on the set of all statements defining the cause of death of a person, and true on the set of all statements defining the cause of death of a person, except the statement A, then we get an example of an identification problem for a system of classical logic leading to a false conclusion. Let's formalize the given example. Let $\Re = \{A_i | i \in I\}$ de a set of all statements, which determine the cause of a person death. We construct the predicate $P: \Re \rightarrow \{0, 1\}$ as follows:

$P(A_i) = 1$, if $A_i \neq A$, $i \in I$, and $P(A_i) = 0$, if $A_i = A$.

1.4 System's Identification Matrix as an Analogue of the Characteristic Function

Let's construct an analogue of the characteristic function—the system's identification matrix as follows:

Definition 1.3 Let $In_S = \{Q_i | i \in I_S\}$ be a set of unary predicates on the set of all algebraic systems of the same signature Ω(or on the set of all groups), specifying the internal attributes of the S. That is: let.

$Q_i : \mathfrak{R} \rightarrow \{0, 1\}$, $i \in I$; $Q_i : \mathfrak{R} \rightarrow \{0, 1\}$, $i \in I$, $\{S_i | i \in I\}$—be the set of all subsystems of the system S. Then the identification matrix that defines the internal attributes of the system M_1 is defined as follows:
$M_1 = m_{ij}$, where

$$m_{ij} = \begin{cases} 1, & if predicate Q_i is true on the subsystem S_j \\ 0, & otherwise \end{cases}$$

By analogy with Definition 1.3, one can define the identification matrix of the external attributes of the system and the identification matrix of the common (internal and external) attributes of the system.

Definition 1.4 Let $Ou_S = \{Q_j | j \in J_S\}$ be the set of unary predicates on the set of all algebraic systems of the same signature Ω(or on the set of all groups), defining external attributes of the system S. That is:

$Q_j : \mathfrak{R} \rightarrow \{0, 1\}$, $i \in I$; $Q_j : \mathfrak{R} \rightarrow \{0, 1\}$, $j \in J$.. Then the identification matrix that defines the external attributes of the system M_2, is defined as follows:
$M_2 = \|m_{ik}\|$, where

$$m_{ik} = \begin{cases} 1, & if predicate Q_j is true on the subsystem S_k \\ 0, & otherwise \end{cases}$$

and $\{S_k | k \in K\}$ is the set of all subsystems of the system S.

Definition 1.5 Let $M = \|m_{ij}\|$, where

$$m_{ij} = \begin{cases} 1, & if predicate Q_i is true on the subsystem S_j \\ 0, & otherwise \end{cases}$$

Here $\{Q_i | i \in I\}$ is the set of properties (external or internal) of the system. The matrix M is called the system S identification matrix.

The system S internal operating conditions in stationary mode should be optimal. The question arises:

How one can build an analogue of ultraproduct in which one can filter by sets of conditions $\{Q_i | i \in I\}$?

1.5 Systems with Two Subsystems and Two Internal Attributes

Let's consider the system S with two internal attributes Q_1 and Q_2. Then the system S identification matrix M that defines the system S internal attributes has the following form:

$M = ||\alpha_{ij}|| \in Z_2^2$, where $\alpha_{ij} = 1$ or $\alpha_{ij} = 0$, $i, j = 1, 2$, Z_2 is a field of two elements. Herewith

$$\alpha_{ij} = \begin{cases} 1, & if\ predicate\ Q_i\ is\ true\ on\ the\ subsystem\ S_j \\ 0, & othewise \end{cases}$$

$$i, j = 1, 2.$$

Identifier matrices addition:

$$M_1 + M_1' = ||\alpha_{ij} + \alpha_{ij}'||$$

$1 + 1 = 0$. So, properties Q_1 and Q_1' don't distinguish subsystems S_1', S_2'.

1.5.1 Orthogonality. The Subsystem S_j Identifier in the System S. Structural Analogue of the Method of Principal Components

In this section, we construct a structural (by the structure of system attributes) analogue of the econometric method of principal components, developed by the outstanding Russian algebraist D.K. Faddeev, [10]. Let's start by looking at the following questions and an example.

The following questions arise:

How one can interpret the identification matrices multiplication?

Let $\{S_1, S_2\}$ be the set of all system S subsystems of with attributes Q_1, Q_2, and $\{S_1', S_2'\}$ be the set of all system S' subsystems of with attributes Q_1', Q_2'.

$$M_1 = \begin{pmatrix} Q_1(S_1) & Q_1(S_2) \\ Q_2(S_1) & Q_2(S_2) \end{pmatrix},$$

$$M_1' = \begin{pmatrix} Q_1'(S_1') & Q_1'(S_2') \\ Q_2'(S_1') & Q_2'(S_2') \end{pmatrix}.$$

Then

$$M_1 M_2' = \begin{pmatrix} Q_1(S_1)Q_1'(S_1') + Q_1(S_2)Q_2'(S_1') & Q_1(S_1)Q_1'(S_2') + Q_1(S_2)Q_2'(S_2') \\ Q_2(S_1)Q_1'(S_1') + Q_1(S_2)Q_2'(S_1') & Q_2(S_1)Q_1'(S_2') + Q_2(S_2)Q_2'(S_2') \end{pmatrix}$$

If the product of the matrices $M_1 M_1' = O$, where O is the zero square $2 \times 2-$ matrix over the field of two elements Z_2, then this means that the matrices M_1, M_1' are orthogonal. Then following question arises:

How one can find the orthogonal basis of the space of identifying matrices system S?

Definition 1.6 The identifier of the subsystem S_j in the system S is called a row vector.

$$\bar{s} = \langle Q_k(S_i) | k = 1, \dots, n \rangle.$$

The identifier of the property Q_i in the system S is called a row vector

$$\bar{q} = \langle Q_i(S_k) | k = 1, \dots, n \rangle.$$

We assume that the number of system's properties coincides with the number of system's subsystems. If this is not so, then one can equalize the corresponding indicators, zeroing the missing coordinates of the vector.

Let the vectors $\bar{q}$ and $\bar{s}$ be orthogonal, that is, their scalar product is equal to 0:

$$\bar{qs} = \sum_{k=1}^{n} Q_i(S_k) Q_k(S_i) = 0$$

In the vector space Z_2^n over the field of two elements Z_2 we apply the orthogonalization process to the system of vectors $\langle Q_k(S_i) | k = 1, \dots, n \rangle, i = 1, \dots, n$, and thus obtain the orthogonal basis of the subsystems of the system S. Applying the orthogonalization process to the system of vectors $\langle Q_i(S_k) | k = 1, \dots, n \rangle, i = 1, \dots, n$, of the space Z_2^n over the field of two elements Z_2, we obtain the orthogonal basis of the properties of the system S.

Here from we obtain the definitions of generating or, better saying, basic subsystems of the system S, basic external attributes of the system S, basic internal attributes of the system S, basic attributes of the system S.

1.6   Decomposition of the System by Its Basic Subsystems and Property. Decomposition of Attributes of a System by Its Basic Attributes and a Subsystem

Now we shall introduce the following definitions that will allow one to decompose the system by the selected property of the system and by its subsystems, and by its selected subsystem and a fixed property, using the results of previous section. It should be noted right away that decomposing a system by its basic subsystems and a property and decomposing a system by its attributes and a subsystem do not imply the ability to synthesize a system from basic subsystems.

Definition 1.7 The basic subsystems of the system S by its external attribute Q_i (property) of the system S are such its subsystems $S_1, S_2, \ldots, S_r$, that the vectors system $\langle Q_i(S_k)|k = 1, \ldots, r \rangle$, forms an orthogonal basis of the space Z_2^n over the field of two elements $Z_2,, i = 1, \ldots, n$.

Definition 1.8 The basic external attributes (properties) of the system S by its subsystem S_i are such its external attributes $Q_1, Q_2, \ldots, Q_m$, for which the vectors system $\langle Q_k(S_i)|k = 1, \ldots, m \rangle$, forms an orthogonal basis of the space Z_2^n over a two-element field $Z_2, i = 1, \ldots, n$.

Definition 1.9 The basic internal attributes (properties) of the system S by its subsystem S_i are such its internal attributes $Q_1, Q_2, \ldots, Q_m$, for which the vectors system $\langle Q_k(S_i)|k = 1, \ldots, m \rangle$, forms an orthogonal basis of the space Z_2^n over a two-element field $Z_2, i = 1, \ldots, n$.

Definition 1.10 The basic attributes (properties) of the system S by its subsystem S_i are such its attributes $Q_1, Q_2, \ldots, Q_m$, for which the vectors system $\langle Q_k(S_i)|k = 1, \ldots, m \rangle$, forms an orthogonal basis of the space Z_2^n over a two-element field Z_2, $i = 1, \ldots, n$.

As for examples, we shall consider models with one basis subsystem S_i of a system S and one defining property in the next section. Such models can be used for systems with poorly known statistics.

1.7   Probability Identification Matrices

In this section, we shall show how the identification matrices of the system can be used in practice. Let's begin with definitions.

Let $M_F = ||F(Q_i(S_j))||, i = 1, \ldots, m, j = 1, \ldots, n$, where $F(Q_i(S_j))$ is the distribution function of the predicate (property) Q_i on the subsystem S_j of the system S.

If one considers $Q_i(S_j)$ as a random variable taking possible values 0 and 1 on the subsystem S_j of the smart system S—then we get Bernoulli distribution on the

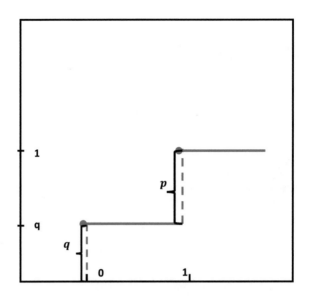

Fig. 1.4 The distribution function for binomial distribution

subsystems S_j of the system S, if j is fixed and i changes. So, we have a random finite vector which consists of 0 and 1 (Fig. 1.4).

It is also possible to fix i, i.e. property Q_i, and j, i.e. subsystem S_j, to change.

By the matrix rank theorem, the rank of the matrix $r(M_F)$ is equal to the largest minor order of the submatrix M_F, which is nonzero. This minor $M_{\alpha\beta}$ gives the maximum (in terms of the number of elements) system of linearly independent functions of vectors $\{F(Q_i(S_j)|j = 1, \ldots, k)\}, i = 1, \ldots, k$, which form the basis of the matrix M_F.

Definition 1.11 The system of functions $\{F(Q_i(S_j)|j = 1, \ldots, k)\}, i = 1, \ldots, k$, is called the probabilistic basis of the system S.

To use this technique in practice, one should empirically try to select statistics for the distribution functions $F(Q_i(S_j))$, and from this we obtain the basic attributes of the system S, the basic subsystems of the system S, the basic external attributes of the system S, the basic internal attributes of the system S.

In particular, one obtains that if $M_1 = \|F(Q_i(S_j))\| \in R^{n \times n}$ and $r(M_1) = n$, then all subsystems $S_j, j = 1, \ldots, n$, are basic, and all attributes $Q_i, i = 1, \ldots, n$, are basic. Applying the orthogonalization process to the systems of vectors $\langle F(Q_i(S_k))|k = 1, \ldots, n\rangle, i = 1, \ldots, n$, of the space Z_2^n over the field of two elements Z_2, we obtain the orthogonal basis of the properties of the system S.

Fig. 1.5 A lattice of
subgroups of the a
sixth-order cyclic group Z_6

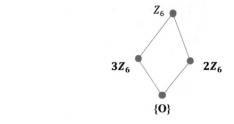

Fig. 1.6 A lattice of
subgroups of the a
sixth-order a third-degree
permutation group S_3

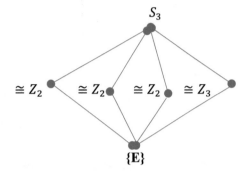

1.8 Examples of System Models with the Same Set of Basis Subsystems and Defining Attributes, the Same Matrix $M = ||F(Q_i(S_j))||$ and Different Structures

For example, we consider a model that is a sixth-order cyclic group Z_6 and a model that is a third-degree permutation group S_3. Both models are of the order six, but their structures differ from each other. We represent the lattices of the subgroups Z_6 и S_3 (Figs. 1.5 and 1.6).

References

1. Home page, https://en.wikipedia.org/wiki/System_identification
2. Home page, https://en.wikipedia.org/wiki/Pattern_recognition
3. Serdyukova, N., Serdyukov, V.: Algebraic formalization of smart systems. In: Theory and Practice, Smart Innovation, Systems and Technologies, SIST, vol. 91, Springer Nature, Switzerland, (2018)
4. Serdyukova, N.A., Serdyukov, V.I, Neustroev, S.S., Shishkina, S.I.: Assessing the reliability of the automated knowledge control results. In: 2019 IEEE Glob. Eng. Conf. EDUCON (2019)
5. Serdyukov, V.I.: On the quantitative estimation of the reliability of the results of the automated control of knowledge. Inf. Educ. **3**, 1–5 (2010) (in Russian)
6. Kozlov, O.A., Kurakin, A.S., Serdyukov, V.I.: On automation of training and control of knowledge of the operators of the information system of aerospace search and rescue. Pedagogical Inform. **5**, 14–17 (2011)

7. Serdyukov, V.I., Kurakin, A.S.: Probabilistic approach to evaluation of students ' knowledge monitoring results. Pedagogical Inform. **6**, 31–38 (2011)
8. Serdyukova, N., Serdyukov, V., Neustroev, S.: Testing as a feedback in a smart university and as a component of the identification of smart systems. In: Uskov, V., Howlett, R., Jain, L. (eds.) Smart Innovation, Systems and Technologies, SIST, vol. 144, pp. 527–538. Springer (2019)
9. Fucks, L.: Infinite abelian groups, vol. 2, p. 416. Academic Press, New York and London (1973)
10. Dubrov, A.M., Mkhitaryan, V.S., Troshin, L.I.: Multidimensional statistical methods and fundamentals of econometrics. Moscow State University of Economics, Stat. Inform. Moscow (2002).((in Russian))

Chapter 2
The Complexity of the Identification Problem. Transition from by Element-Wise Model of the System to the Model of Factors Determining the System. Quasi-fractal Algebraic Systems

However, in mathematics there are no coincidences.
Mathematics is the realm of necessity.
Even the most superficial and casual in appearance
coincidence always expresses a deeper analogy.
N. Vavilov, Concrete group theory.

Abstract In this chapter the following basic questions are considered:

- models' presentation forms,
- statistics methods of system identification,
- algebraic methods of systems identification,
- transition from the element-wise model of the system to the model of factors determining the system,
- fractal forms of model representation and systems identification methods,
- quasi-fractal algebraic systems

Keywords Identification problem · Quasi-fractal system

2.1 Introduction

Main idea of chapter concerning the problem of representation of a model of a system, runs as follows: we would like to find such a form of a model for a smart system that one can refine its properties. The quasi-fractal systems allow one to use the concept of algebraic formalization of smart systems for a more in-depth study of smart systems, using the possibilities of its study, being at different levels of smart system's detail.

Advantages of the algebraic quasi-fractal form of representation of models are as follows.

The quasi-fractal form of representing the models of the system S in the algebraic formalization of systems allows us to consider major, initially distinguished factors that determine the system S, in the form of algebraic systems, that is, it allows, in

© Springer Nature Switzerland AG 2021

N. A. Serdyukova and V. I. Serdyukov, *Algebraic Identification of Smart Systems*, Intelligent Systems Reference Library 191, https://doi.org/10.1007/978-3-030-54470-6_2

fact, to refine the model G_S of factors determining the system S, onto any level $k = 1, \ldots n, \ldots$.

The complexity of solving the problem of identification of systems is also in the variety of forms of representation of models of systems. Let's focus on some forms of model representation.

2.2 Models' Presentation Forms

Traditionally, the following forms of model representation are used:

- In the form of equations—mainly in vector spaces. In this case, in fact, the following algebraic formalization of the model is used: the algebraic system $V = \langle V, +-, \{\lambda \in P\} \rangle$, which is a vector space over the field P.
- In the form of arbitrary algebraic systems [1].
- Through fractals [2]. In [2] there is an example of a simple game-theoretic model in which many optimal wins have a pronounced fractal structure.
- We shall introduce the concept of a quasi-fractal algebraic system, a quasi-fractal algebraic system of level k, $k = 1, \ldots, n, \ldots$ and we shall show that any algebraic system is a first-level quasi-fractal.

2.3 Statistics Methods of System Identification

The model for solving the statistical identification problem is usually presented in the form of an equation, usually it is an equation in vector space. Thus, when using statistical identification, only part of the model, namely its function (main or target) is visible [3]. The rest of the model is hidden and not considered. Under algebraic formalization, a model is an algebraic system. In a Euclidean finite-dimensional space, a metric is used for calculations; in other algebras, a measure can be used.

2.4 Algebraic Methods of Systems Identification. Transition from the Element-Wise Model of the System to the Model of Factors Determining the System

The complexity of representing the model using the element-wise approach (embedding in a holomorph of a group) explains the difficulties in solving the problem of identification when using element-wise models of the system.

In this regard, the following question arises: how to describe the inverse transition from the element-wise model of the system to the model of factors that determine the system? One can solve this problem using Chap. 6 of the book [1]. To do this it is necessary to build identification matrices of the factor model and the model consisting of elements of the system—the element-wise model of the system. The inverse transition from the element-wise model of the system to the model of factors that determine the system raises the question: how to find a group by the holomorph of a group that has the same holomorph (a given holomorph)? To do this, one has to consider holomorphically isomorphic groups. While modeling systems with permutation factors that determine the system, for example, if the model of the system is an Abelian group of factors that determine the system (for models of systems with permutation factors), for example, in the case of ranking systems in education, one can use the works of the Tomsk school of mathematics. This leads to subsystems for evaluating control systems. Further models can be classified according to the areas in which they operate.

2.5   Fractal Forms of Model Representation and Systems Identification Methods

In the work [4], it is noted that one of the largest sections of fractal geometry is the theory of self-similar fractals, which originates from the article by J. Hutchinson "Fractals and Self Similarity" (1981) and which has now turned into independent section of mathematics [5]. Hutchinson introduced the concept of an invariant set in a complete metric space as a compact set made up of its images under the action of some finite set of contraction mappings of a given metric space into itself. Such sets of contraction mappings are usually called iterable functions systems (IFS), and their invariant sets are called attractors. An attractor of a system of contraction similarities is called a self-similar set, and, similarly, an attractor of a system of contraction affine mappings in Banach space is called a self-affine set. Fractal geometry methods are widely used in mathematical modeling. For a more detailed study of IFS attractors, Hutchinson proposed to consider measures on fractals and introduced invariant measures of a system of contraction maps, also called self-similar measures.

Example   When we read a text and our brain processes the information we read, we look for the essence of it, that is, we compress the information. Then we do the same again and again…. In fact, we are looking for a fixed point of the compressive map by Brouwer's fixed point theorem. This seems to be the essence of the discretization of mathematics: from continuity we pass to certain points-fixed points of contraction maps and thus seek the essence of the described processes and phenomena. The reverse process that is the transition from continuous mathematics to discrete is an expanding process.

2.6   Quasi-fractal Algebraic Systems

Fractal models can be considered as a generalization of classical models. In this case, the usual or classical model is a quasi-fractal of the first level.

Definition 2.1 Let's consider the algebraic system $A_1 = \langle A_1; \Omega_1 \rangle$ of the signature Ω_1, such that every element a_α, $\alpha \in \Lambda_1$, of the main set A_1 of the system A_1 in turn is an algebraic system of the signature Ω_2. That is $a_\alpha = A_2^\alpha = \langle A_2^\alpha; \Omega_2 \rangle$, $\alpha \in \Lambda_1$ is an algebraic system of the second level. Continue this process by induction. If an algebraic system $A_k^\alpha = \langle A_k^\alpha; \Omega_k \rangle$, $\alpha \in \Lambda_k$ is an algebraic system of the level k of the fractal and every element a_α, $\alpha \in \Lambda_k$ of the main set A_k^α of the system A_k^α is an algebraic system $a_\alpha = A_{k+1}^\alpha = \langle A_{k+1}^\alpha; \Omega_{k+1} \rangle$ of the signature Ω_{k+1}, of the level $k + 1$, $\alpha \in \Lambda_{k+1}$, of the fractal, then the algebraic system $A_1 = \langle A_1; \Omega_1 \rangle$ is called a quasi—fractal algebraic system. If all signatures Ω_k, $k = 1, \ldots n, \ldots$, are equal to each other and all the systems $\langle A_k^\alpha; \Omega_k \rangle$, $\alpha \in \Lambda_k$, are isomorphic to each other then the algebraic system $A_1 = \langle A_1; \Omega_1 \rangle$ is called a fractal algebraic of the signature Ω_1.

We should explain that in this notation $A_1 = \langle A_1; \Omega_1 \rangle$ plays the role of a universal variable to denote a quasi-fractal algebraic system.

Let's note that an ordinary algebraic system is a quasi-fractal system of the first level.

So, we have some advantages of the algebraic quasi-fractal form of representation of models.

The quasi-fractal form of representing the models of the system S in the algebraic formalization of systems allows us to consider major, initially distinguished factors that determine the system S, in the form of algebraic systems, that is, it allows, in fact, to refine the model G_S of factors determining the system S, onto any level $k = 1, \ldots n, \ldots$.

2.6.1   Contraction Mappings of a Quasi-fractal Algebraic System

According to [6], the first of the concepts, concerning self-organized criticality and characterizing the complexity of the system under study is scale invariance, which means that events or objects lack their own characteristic sizes, durations, energies, etc. Scale invariant systems are arranged identically at all levels of organization, i.e. they do not have a scale that would be responsible for the "most important processes". Systems that are arranged identically at all levels of self-organization remind fractals. In turn, the presence of scales at different levels of the system under study is a sign of the simplicity of the system, and usually serves as a condition for the application of traditional methods of mathematical modeling. Fractal mathematical models are characterized by compression processes. In fact, in algebraic quasi-fractals there is a process of "qualitative compression" when moving across the fractal levels from top to bottom. At the same time, one can also build compressive quasi—fractal mappings and compress quasi—fractal levels in different ways. Let's consider some examples.

Example 2.1 We can construct contraction mappings of a quasi-fractal algebraic system going down from top to bottom along the levels of the quasi-fractal according to the following scheme:

See Fig. 2.1.

Example 2.2 We can construct contraction mappings of a quasi-fractal algebraic system going down from top to bottom along the levels of the quasi-fractal according to the following scheme:

See Fig. 2.2.

Example 2.3 We can construct contraction mappings of a quasi-fractal algebraic system going down from top to bottom along the levels of the quasi-fractal according to the following scheme:

See Fig. 2.3.
Now let's describe some contraction mappings in metric spaces.

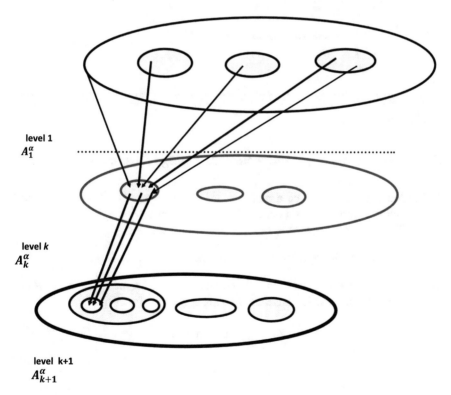

level 1
A_1^α

level k
A_k^α

level k+1
A_{k+1}^α

Fig. 2.1 For Example 1

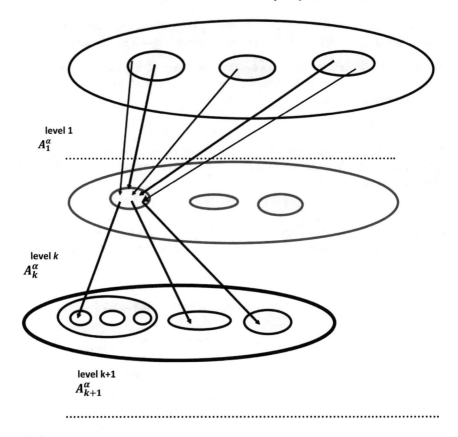

level 1
A_1^α

level k
A_k^α

level $k+1$
A_{k+1}^α

Fig. 2.2 For Example 2

2.6.2 Metric Spaces. Contraction Mappings Examples

Definition 2.2 Let $\langle X, d \rangle$ be a metric space. A mapping $T : X \to X$ is called a contraction mapping with a compression coefficient $s, 0 < s < 1$, if for every $x, y \in T$ the inequality $d(T(x), T(y)) \le sd(x, y)$ is true.

Also recall now the definition of a metric space.

Definition 2.3 Metric space is an ordered pair $\langle X, d \rangle$, where $d : X \times X \to R$ is a non-negative function satisfying the following conditions for every $x, y, z \in X$:

$$d(x, y) = 0 \Leftrightarrow x = y$$

$$d(x, y) = d(y, x)$$

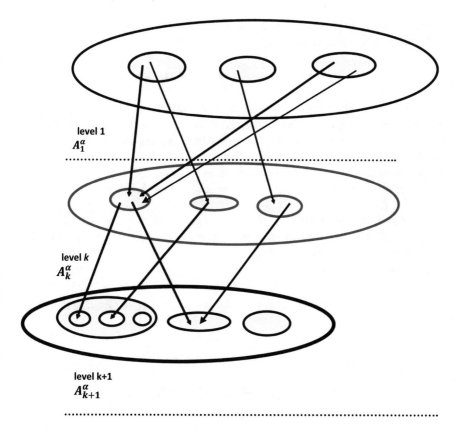

level 1
A_1^α

level k
A_k^α

level k+1
A_{k+1}^α

Fig. 2.3 For Example 3

$$d(x, z) \leq d(x, y) + d(y, z)$$

A map $T : X \to X$ is called a weakly contraction map if the following inequality

$$d(f(x), f(y)) < d(x, y)$$

holds for any different points $x, y \in X$.

It is known that any weakly contraction map of a compact to itself has a single fixed point [7].

Example 2.4 Let's define a metric on the quasi-fractal algebraic system $A_1 = \langle A_1; \Omega_1 \rangle$ in the following way.

Definition 2.4 Let's define $d : \langle A_1; \Omega_1 \rangle \times \langle A_1; \Omega_1 \rangle \to R$ in the following way:

let $d\left(A_x^\infty, A_y^\infty\right) = |y - x|$.
Then as $|z - x| \leq |y - x| + |z - y|$, conditions 1–3 take place.

Now let's consider some examples of contraction mappings.

Let's construct contraction mapping

$$T : \boldsymbol{A}_1 = \langle A_1; \Omega_1 \rangle \rightarrow \langle \boldsymbol{A}_1 = A_1; \Omega_1 \rangle$$

Let $T\left(A_x^{\propto}\right) = A_{sx}^{\propto}$, where, $0 < s < 1$. If sx is an integer, then $\left(A_{sx}^{\propto}, \Omega_{sx}\right)$ is situated on the sx level of a quasi—fractal algebraic $\boldsymbol{A}_1 = A_1; \Omega_1$. Let $s = \frac{1}{m}, m \in N$. Let's consider a one-to-one mapping $\theta : N \rightarrowtail mN$ It defines a one-to-one mapping T of a quasi—fractal $\boldsymbol{A}_1 = A_1; \Omega_1$ onto itself, but at the same time, all levels of the image of the quasi-fractal $\boldsymbol{A}_1 = A_1; \Omega_1$ will be multiples of m.

For levels which are multiples of m we have: $m \mapsto 1, 2m \mapsto 2, \ldots, km \mapsto k, \ldots, k \in N$.

For other levels: if for a level k we have: $k = ma + b, 0 \le b < m, k > m$, then let $T\left(A_k^{\propto}\right) = A_a^{\propto}$. So, we have $ma + a > ma$ and the compression condition is satisfied. The first m levels are not compressed.

Now, let

$$d\left(A_1, A_x^{\propto}\right) = \frac{1}{x} \quad \text{for any level } x \text{ of a quasi - fractal } \boldsymbol{A}_1 = \langle A_1; \Omega_1 \rangle \qquad (2.1a)$$

Then we have:
$\frac{1}{x+y} \le \frac{1}{x} + \frac{1}{Y} \Leftrightarrow \frac{1}{x+y} \le \frac{x+y}{xy}$—obvious inequality, that is:

$$d\left(A_1, A_{x+y}^{\propto}\right) \le d\left(A_1, A_x^{\propto}\right) + d\left(A_1, A_y^{\propto}\right)$$

$$d\left(A_x^{\propto}, A_y^{\propto}\right) = \begin{cases} \frac{1}{|y-x|}, & x \ne y \\ 0, & x = y \end{cases}$$

Conditions 1–3 are satisfied with this definition of function d.
To prove this, it suffices to prove the inequality:

$$\frac{1}{|z - x|} \le \frac{1}{|y - x|} + \frac{1}{|z - y|}$$

Example 2.5 Now let

$$d\left(A_x^{\propto}, A_y^{\propto}\right) = \begin{cases} \frac{1}{x} + \frac{1}{y}, & x \ne y \\ 0, & x = y \end{cases} \qquad (2.2a)$$

Let's check out the fulfillment of the third condition of the definition 2.3. Really

$$\frac{1}{x} + \frac{1}{z} \le \frac{1}{x} + \frac{1}{y} + \frac{1}{y} + \frac{1}{z}.$$

Therefore (2.2a) defines a metric on an algebraic quasi-fractal $\boldsymbol{A}_1 = \langle A_1; \Omega_1 \rangle$.

Let's define in this case the compression ratio s: $\frac{1}{k} - \frac{1}{k+1} = \frac{1}{k(k+1)}$, that is $s = \frac{1}{k}$ is the compression ratio between the levels k and $k + 1$.

That is, if one looks for a parameter s suitable for all levels of the system $A_1 = A_1; \Omega_1$ which are greater than the first level, then we can put $s = \frac{1}{2}$.

Example 2.6 It is known that if $\langle X, d \rangle$ is a metric space, then $\langle X, \frac{d}{d+1} \rangle$ is also a metric space, and the metrics d and $\frac{d}{d+1}$ induce the same topology on X. Therefore, $\langle A_1, \frac{1}{x+y+1} \rangle$ is also a metric space with the same topology as in Example 2.5.

Hutchinson [5], introduced the concept of an invariant set in a complete metric space as a compact set composed of its images under the action of some finite set of contraction mappings of a given metric space into itself. Such sets of contraction mappings are called systems of iterable functions (IFS), and their invariant sets are called attractors. An attractor of a system of contraction similarities is called a self-similar set. Let's recall now the Banach fixed—point theorem. Banach's theorem has many different formulations and interpretations.

Recall that a metric compact is a metric space that is compact as a topological space with a topology induced by the metric. A point $a \in X$ is called a fixed point of the map $f : X \rightarrow X$, if $f(a) = a$.

The fixed point of the contraction mapping $f : A_1 \rightarrow A_1$ allows one to determine the optimal level k of the quasi-fractal A_1 the level of its predictability.

Theorem (Picard-Banach fixed point principle) The contraction map $f : X \rightarrow X$ of the complete metric space into itself has a fixed point a, more over this fixed point is a single one. Moreover, for any point $x_0 \in X$ the iterative sequence

$$x_0, x_1 = f(x_0), \ldots, x_{n+1} = f(x_n), \ldots$$

converges to a. The rate of this convergence is given by the estimate $d(a, x_n) \leq \frac{s^n}{1-s} d(x_1, x_0)$.

The points $x_0, x_1 = f(x_0), \ldots, x_{n+1} = f(x_n), \ldots$ are called successive approximations to a fixed point. In [7] it is shown how one can look at the fixed point theorem in a different way. Namely, it is noted that the following Tarski theorem on the monotone mapping of the complete lattice into itself is closely related to the fixed point theorem. Examples of a complete lattice are any segment $[a, b] \subset R$, and the set of all subsets of an arbitrary nonempty set.

Tarski's Theorem Let $f : X \rightarrow X$ be a monotone mapping of a complete lattice X into itself. Then f has a fixed point.

Also, the fixed-point theorem is closely related to Brouwer's theorem. Recall its wording.

Brouwer's fixed-point theorem
If $f : X \rightarrow X$ is a continuous map of a convex compact X into itself, then there exists a fixed point of the map f.

In [7], it was shown that any convex compact set of dimension n is homeomorphic to the unit ball D_n in R^n, where $D_n = \{x \in R^n \,||x| \leq 1\}$ and $|x| = |x_1, \ldots, x_n| = \left(x_1^2, \ldots, x_n^2\right)^{\frac{1}{2}}$.

Main Theorem

Let the system S be modeled by a quasi-fractal algebraic system $A_1 = \langle A_1; \Omega_1 \rangle$. Then the system S has a finite level of forecasting (predictability).

Proof Let's define the metric d on the quasi-fractal $A_1 = \langle A_1; \Omega_1 \rangle$ in accordance with Example 2.4. Then we embed the metric quasi—fractal $A_1 = \langle A_1; \Omega_1 \rangle$ into its completion by this metric, which is a metric compact with metric d. According to [7], this is a complete metric compact, in the case where its dimension n, is homeomorphic to the unit ball D_n in R^n, where $D_n = \{x \in R^n \,||x| \leq 1\}$ and $|x| = |x_1, \ldots, x_n| = \left(x_1^2, \ldots, x_n^2\right)^{\frac{1}{2}}$. Then, according to Brouwer's theorem, we obtain a fixed point of the contraction map from Example 2.4, which sets the level of predictability of the system S.

2.7 Algebraic Fractals

In this section, we shall consider questions about how, using an algebraic fractal system, that is, using an algebraic quasi-fractal, one can write the following well-known constructions of algebra and category theory:

– free product,
– Cartesian product,
– direct sum.

Definition 2.5 An algebraic fractal is a quasi-fractal algebraic system.

2.7.1 Product and Coproduct Record of Category Theory in the Form of a Quasi-fractal Algebraic System

Let's recall the definition of a product adopted in category theory

Definition 2.6 Let the set $\{X_i | i \in I\}$ of an indexed family of (not necessarily distinct) category objects be given C. An object X of a category C together with the family of morphisms $\pi_i : X \to X_i, i \in I$, is the product of a family of objects $\{X_i | i \in I\}$ if for any object $Y \in Ob(C)$ and any family of morphisms $f_i : Y \to X_i, i \in I$, there exists a single morphism $f : Y \to X$ for which the following diagram:

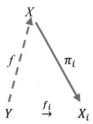

is commutative for each $i \in I$, that is $\pi_i f = f_i$. In this case, for the product the designation is used: $X = \prod_{i \in I} X_i$.

Let's recall that in the category of sets, the category product coincides with the Cartesian product. In the category of topological spaces, the product of spaces corresponds to a space whose support is the Cartesian product of the supports of the factors, and the topology is defined as the product of their topologies. In the category of groups, the product of groups is defined as their direct product. In the category of projective varieties, a category product can be defined using the Segre embedding. A partially ordered set can be considered as a category in which a morphism from a to b exists if and only if (by definition) $a \geq b$ (moreover, there cannot be more than one morphism between two objects). In this case, the product of the family of linearly ordered objects is their largest lower bound, and coproduct is the smallest upper bound.

2.7.2 Record (Cartesian) Product of a Countable Number of Algebraic Systems in the Form of a Quasi-fractal Algebraic System (in the Form of an Algebraic Fractal)

Definition 2.7 Let there be a nonempty set of algebraic systems $X_i = \langle X_i; \Omega_i \rangle$, $i = 1, \ldots, n, \ldots$, contained in some category of algebraic systems with products [1]. Then every quasi—fractal $A_1 = \langle A_1; \Omega_1 \rangle$, $A_1 = \{ A_1^1, A_2^1, \ldots, A_n^1, \ldots \}$, where $X_1 = \langle X_1; \Omega_1 \rangle$ is the first element of the set A_1, that is $X_1 = \langle X_1; \Omega_1 \rangle = A_1^1$, $X_2 = \langle X_2; \Omega_2 \rangle$ is the second element of the set A_2^1, that is $X_2 = X_2; \Omega_2 = A_2^1, \ldots, X_i = \langle X_i; \Omega_i \rangle$ is the i th element of the set A_i^1, that is $X_i = \langle X_i; \Omega_i \rangle = A_i^1, \ldots$ and so on,

---

[1] There are categories in which there are finite families of objects that do not have a product, for example, the category of ordered Abelian groups [8]. Subcategories of the categories of all sets, as well as categories that are isomorphic to them, are called concrete categories. The main examples of categories, such as categories of groups, topological spaces, etc., belong to concrete categories; indeed, their objects—groups, topological spaces, etc.—are sets, and mappings are homomorphisms, continuous mappings, etc., are some unique (univalent) mappings of these sets [9]. In the category of all groups, the categorical concept of direct union coincides with the concept of a complete direct product of groups, and therefore for finite families of groups it coincides with the concept of a direct product.

can be considered as quasi—fractal which contains a cartesian product of algebraic systems $X_i = \langle X_i; \Omega_i \rangle, i = 1, \ldots, n, \ldots$.

So, quasi-fractal $A_1 = \langle A_1; \Omega_1 \rangle$ contains a cartesian product of algebraic systems $X_i = \langle X_i; \Omega_i \rangle, i = 1, \ldots, n, \ldots$.

Besides it, if one puts $f = f_i \varphi_i, i = 1, \ldots, n, \ldots$ where $\varphi_i, i = 1, \ldots, n, \ldots$ is an isomorphism $\varphi_i : X_i \to A_i^1$, then the following diagram

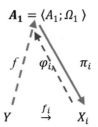

is a commutative one. That is, $A_1 = \langle A_1; \Omega_1 \rangle = \prod_{i=1,\ldots,n,\ldots} (X_i = \langle X_i; \Omega_i \rangle)$

So, X_i is written as an element with the number i of the quasi-fractal $A_1 = \langle A_1; \Omega_1 \langle$, standing at the first level of the fractal.

Remark

(1) The quasi-fractal system $A_1 = \langle A_1; \Omega_1 \rangle$ does not belong to the category of algebraic systems generated by algebraic systems $X_i = \langle X_i; \Omega_i \rangle, i = 1, \ldots, n, \ldots$. Therefore, the uniqueness theorem of a Cartesian product [10], does not work here.

(2) Representation of the product $\prod_{i=1,\ldots,n,\ldots} (X_i = \langle X_i; \Omega_i \rangle)$ as a quasi-fractal (Definition 2.7) is not the only possible.

2.7.3 Record Coproduct of a Countable Number of Algebraic Systems in the Form of a Quasi-fractal Algebraic System (in the Form of an Algebraic Fractal)

Definition 2.8 An object A together with a family of morphisms $u_i : A_i \to A, i \in I$, is called a coproduct (sum) of the family of objects $A_i, i \in I$, if for every object X of category C the map $Hom_C(A, X) \to \prod_{i \in I} Hom_C(A_i, X)$, associating with each morphism $u : A \to X$ the family of morphisms $(uu_i), i \in I$, is bijective.

Let there be a nonempty set of algebraic systems $X_i = \langle X_i; \Omega_i \rangle, i = 1, \ldots, n, \ldots$, which is contained in a certain category of algebraic systems with products[2]. Then any quasi-fractal $A_1 = \langle A_1; \Omega_1 \rangle, A_1 = \{A_1^1, A_2^1, \ldots, A_n^1, \ldots\}$,

---

[2]There are categories in which there are finite families of objects that do not have a product, for example, the category of ordered abelian groups [8]. Subcategories of the categories of all

where $X_1 = \langle X_1; \Omega_1 \rangle$ is the first element of the set A_1, that is $X_1 = \langle X_1; \Omega_1 \rangle = A_1^1$, $X_2 = \langle X_2; \Omega_2 \rangle$ is the second element of the set A_2^1, i.e. $X_2 = \langle X_2; \Omega_2 \rangle = A_2^1$,..., $X_i = \langle X_i; \Omega_i \rangle$ is the i-th element of the set A_i^1, that is, $X_i = \langle X_i; \Omega_i \rangle = A_i^1$,..., and so on, can be considered as quasi—fractal which contains a coproduct of algebraic systems $X_i = \langle X_i; \Omega_i \rangle, i = 1, \ldots, n, \ldots$.

So, quasi-fractal $A_1 = \langle A_1; \Omega_1 \rangle$ contains a coproduct of algebraic systems $X_i = \langle X_i; \Omega_i \rangle, \quad i = 1, \ldots, n, \ldots$.

2.7.4 Special Case. Quasi-fractal Groups

We show that a quasi-fractal group allows us to represent the coproduct of a countable number of groups in the form of a 2×2 construction (flat construction), ($m \times i$— dimensional construction), where m is the element number in the word, i is the level of the quasi-fractal system, $i = 1, \ldots, n, \ldots$ and not one-dimensional. That is, we introduce the Cartesian coordinate system xOy, on the plane, on the Ox axis we shall put off the number of the element the word, on the Oy axis we will put off the level in the quasi-fractal group.

In fact, let's consider the case when all $X_i = \langle X_i; \Omega_i \rangle, i = 1, \ldots, n, \ldots$, are groups. In the groups category, the coproduct coincides with the free product. Let $w_1 = w_1(X_i | i \in I)$ be a word from $X_i = \langle X_i; \Omega_i \rangle, i = 1, \ldots, n, \ldots, w_2 = w_2(X_i | i \in I)$ be a word from $X_i = \langle X_i; \Omega_i \rangle, i = 1, \ldots, n, \ldots$, and the group operation on the set of all words in the alphabet $\left\{ X_i; \Omega_i \cup X_i^{-1}; \Omega_i \cup \emptyset, i = 1, \ldots, n, \ldots \right\}$ is a sequential writing of words, (see Theorem 2.10 below). Then the first element of the word w_1 is the element in the first level of the quasi-fractal $A_1 = \langle A_1; \Omega_1 \rangle$, which stands in the place corresponding to this element in the word (if, for example, it is X_m, then it stands in place with number m on the first level). The second element of the word w_1 is the element in the second level of the quasi-fractal $A_1 = \langle A_1; \Omega_1 \rangle$ which stands in the place corresponding to this element in the word (if, for example, it is X_m, then it stands in place with number m on the second level). And so on, that is we continue to write sequentially out the elements of the word w_1. We continue to write out the elements of the word w_2 starting from the level $||w_1||$, where $||w_1||$ is word length. Then the fractal $A_1 = \langle A_1; \Omega_1 \rangle$ contains the coproduct $\coprod\limits_{i \in I} X_i$ of groups $X_i = \langle X_i; \Omega_i \rangle, i = 1, \ldots, n, \ldots$.

---

sets, as well as categories that are isomorphic to them, are called concrete categories. The main examples of categories, such as categories of groups, topological spaces, etc., belong to specific categories; indeed, their objects—groups, topological spaces, etc.—are sets, and mappings are homomorphisms, continuous mappings, etc., are some unique mappings of these sets [9]. In the category of all groups, the categorical concept of direct union coincides with the concept of a complete direct product of groups, and therefore for finite families of groups it coincides with the concept of a direct product.

The quasi-fractal system $A_1 = \langle A_1; \Omega_1 \rangle$ does not belong to the category of groups generated by groups $X_i = X_i; \Omega_i, i = 1, \ldots, n, \ldots$, so the uniqueness theorem of a coproduct [10], does not work here.

Now let's consider the diagrams corresponding to the product and coproduct. In Category Theory, the product corresponds to the following scheme [10].

Product

$$
\begin{array}{c}
B \\
\alpha \searrow \sigma_i \\
A = \prod_{i \in I} A_i \xrightarrow{\pi_i} A_i \to 0
\end{array}
\tag{2.1b}
$$

$$\pi_i \alpha = \sigma_i$$

Coproduct

$$
\begin{array}{c}
1 \to A_i \xrightarrow{\tau_i} \coprod_{i \in I} A_i \\
\alpha_i \downarrow \quad \nearrow \beta \\
B
\end{array}
$$

Let's consider the diagram

$$
A = \prod_{i \in I} A_i \xrightarrow{\pi_i} A_i \xrightarrow{\tau_i} \coprod_{i \in I} A_i
$$
$$
\pi_i \quad \alpha_i \quad \gamma
$$
$$
A = \prod_{i \in I} A_i
\tag{2.2b}
$$

Then γ is a natural morphism of a coproduct $\coprod_{i \in I} A_i$ into a product $\prod_{i \in I} A_i$
Let's consider the diagram

$$A = \prod_{i \in I} A_i \overset{\pi_i}{\to} \quad A_i \overset{\tau_i}{\to} \coprod_{i \in I} A_i$$

$$\alpha \quad \sigma_i \mid \tau_i$$

$$\coprod_{i \in I} A_i \tag{2.3}$$

Then α is a morphism of a coproduct $\coprod_{i \in I} A_i$ into a product $\prod_{i \in I} A_i$. Besides it, $\alpha = (\times_{i \in I}) \sigma_i = (*_{i \in I}) \tau_i$

If in the category C there exists a product $\prod_{i \in I} A_i (\pi_i)$ and a coproduct $\coprod_{i \in I} B_i (\tau_i)$ then for any family of morphisms $\alpha_{ij} : B_j \to A_i, i, j \in I$ the following equality is true:

$$\left(*_{j \in I}\right)\left((\times_{i \in I}) \alpha_{ij}\right) = (\times_{i \in I})\left(\left(*_{j \in I}\right) \alpha_{ij}\right)$$

In the case of the groups category, to compare the notions of product and coproduct, one can use [11]. In [11] a free product of groups is characterized as follows:

Definition 2.9 A group A is called a free product of its subgroups $A_i, i \in I$, if:

(1) A is generated by the subgroups $A_i, i \in I$,
(2) For any homomorphisms $\alpha_i : A_i \to B$ there exists a homomorphism $\gamma : A \to B$, whose restrictions on $A_i, i \in I$, coincide with the homomorphisms α_i for any $i \in I$: $\gamma \restriction \tau_i (A_i) = \alpha_i, i \in I$

$$1 \to A_i \overset{\tau_i}{\to} \coprod_{i \in I} A_i$$

$$\alpha_i \quad \gamma \tag{2.4}$$

$$B$$

This directly implies the existence of a natural epimorphism $\alpha : \coprod_{i \in I} A_i \to \prod_{i \in I} A_i$ of the free product (coproduct) $\coprod_{i \in I} A_i$ onto the direct product (product) $\prod_{i \in I} A_i$ of the same family of groups.

The kernel $Ker\alpha$ of the epimorphism α is called the Cartesian subgroup of the free product.

$$1 \to A_i \xrightarrow{\tau_i} \coprod_{i \in I} A_i$$

$$\gamma \upharpoonright \tau_i(A_i) = \alpha_i \qquad\qquad \alpha \qquad\qquad\qquad (2.5)$$

$$\prod_{i \in I} A_i$$

For diagram (2.5) γ is as it is in diagrams (2.2a) and (2.4)
Then:

(1) The kernel $Ker\alpha$ of the natural homomorphism α of the free product of the groups $\coprod_{i \in I} A_i$ onto their direct product $\prod_{i \in I} A_i$ is a free group, that is, the Cartesian subgroup of the free product of groups is free.
(2) $Ker\alpha \cap A_i = \{1\}, i \in I$.

In addition, in [11] the following theorem is proved:

Theorem 2.10 [11]

Let

$$a = a_{i_1} a_{i_2} \ldots a_{i_m} c \qquad\qquad (2.6)$$

where $a_{i_k} \neq 1$ for any k, $i_1 < i_2 < \cdots < i_m, c \in Ker\alpha$. Elements $a_{i_1}, a_{i_2}, \ldots, a_{i_m}, c$ are uniquely determined by the element a and the choice of order on the set I.

Proof Representation (2.6) follows from the fact that each element $a \neq 1$ of the group $\coprod_{i \in I} A_i$ is uniquely represented in the form

$$a = a_1 a_2 \ldots a_l, 1 \neq a_t \in A_{i(t)}, i(t) \neq i(t+1)$$

by a permutation of the syllables. Permutation leads to the appearance of commutators, which can be assembled on the right side with the help of the identity

$$xy = yx[x, y].$$

Using the natural mapping of the coproduct into a product, we obtain that $c \in Ker\alpha$. Let's now use the projection δ_i^* of the coproduct $\coprod_{i \in I} A_i$ onto the group A_i. Then $\delta_{i_k}^*(a) = a_{i_k}$, for any $k = 1, \ldots, m$. This shows that the elements a_{i_k} are uniquely determined by the element a; therefore, c is also determined by the element a, if the order of the elements a_{i_k} is given.

Let's consider representation of quasi-fractal groups on the plane. Suppose now that there is again the xOy plane, along the Ox we shall postpone the number of the element in the word, along the Oy we shall postpone the level in the quasi-fractal group. From definition 2.9 we obtain that the product $\prod_{i=1,\ldots,n,\ldots}(X_i = X_i; \Omega_i)$

is represented by points with integer coordinates on the line $y = 1$. The coproduct $\coprod_{i=1,\ldots,n,\ldots}(X_i = X_i; \Omega_i)$ does not lie on one straight line, as we saw this at the beginning of Sect. 2.7.3.

2.8 Quasi-fractal Graphs

In [12] it is noted that the analysis of a complex network by representing its structure as a fractal with cyclically repeating structures significantly reduces the complexity of solving problems associated with complex structures and systems as a whole. To do this the following technique of replacing elements of the prefractal graph with its subgraphs of the seed is used. A seed is a connected n-vertex graph $H = (W, Q)$ with unmarked vertices [10]. The technique of replacing elements of the prefractal graph with its subgraphs of the seed runs as follows. To determine the oriented fractal (prefractal) graph, we shall need the operation of replacing the vertex with a seed. The essence of this operation is as follows. In this initial graph, a set of adjacent vertices is allocated for the vertex intended for replacement. Next, the vertex and all incident edges are removed from the graph. Then each vertex is connected by an edge to one of the seed vertices. Vertices are connected arbitrarily (randomly) or by a certain rule. The established rules for connecting subgraphs of seeds using old edges allow us to talk about the presence (or absence) of simulated graphs of fractal properties.

Definition 2.11 (see also [13]). Let's consider the graph $\Gamma_1 = \langle\{V_{\alpha_1}^1 | \alpha_1 \in \Lambda_1\}, \{u_{\alpha_1,\beta_1}^1 | \alpha_1, \beta_1 \in \Lambda_1\}\rangle$, where $\{V_{\alpha_1}^1 | \alpha_1 \in \Lambda_1\}$ is the set of vertices of the graph Γ_1, $\{u_{\alpha_1,\beta_1}^1 | \alpha_1, \beta_1 \in \Lambda_1\}$ is the set of edges of the graph Γ_1.

Let's call the graph Γ_1 a graph of the first level. Let, in turn, $V_{\alpha_1}^1$ be the graph $\Gamma_2 = \langle\{V_{\alpha_2}^2 | \alpha_2 \in \Lambda_2\}, \{u_{\alpha_2,\beta_2}^2 | \alpha_2, \beta_2 \in \Lambda_2\}\rangle$, where $\{V_{\alpha_2}^2 | \alpha_2 \in \Lambda_2\}$ is the set of vertices of the graph Γ_2, $\{u_{\alpha_2,\beta_2}^2 | \alpha_2, \beta_2 \in \Lambda_2\}$ is the set of edges of the graph Γ_2. Let's call the graph Γ_2 a graph of the second level. Let's continue this process by induction.

If the graph $\Gamma_k = \langle\{V_{\alpha_k}^k | \alpha_k \in \Lambda_k\}, \{u_{\alpha_k,\beta_k}^k | \alpha_k, \beta_k \in \Lambda_k\}\rangle$, where $\{V_{\alpha_k}^k | \alpha_k \in \Lambda_k\}$ is the set of vertices of the graph Γ_k, $\{u_{\alpha_k,\beta_k}^k | \alpha_k, \beta_k \in \Lambda_k\}$ is the set of edges of the graph—Γ_k, is the graph of the level k, then let $V_{\alpha_k}^k$, $\alpha_k \in \Lambda_k$, be a graph $\Gamma_{k+1} = \langle\{V_{\alpha_{k+1}}^{k+1} | \alpha_{k+1} \in \Lambda_{k+1}\}, \{u_{\alpha_{k+1},\beta_{k+1}}^{k+1} | \alpha_{k+1}, \beta_{k+1} \in \Lambda_{k+1}\}\rangle$, where $\{V_{\alpha_{k+1}}^{k+1} | \alpha_{k+1} \in \Lambda_{k+1}\}$ is a set of vertices of the graph Γ_{k+1}, $\{u_{\alpha_{k+1},\beta_{k+1}}^{k+1} | \alpha_{k+1}, \beta_{k+1} \in \Lambda_{k+1}\}$ is the set of edges of the graph Γ_{k+1}. The graph Γ_{k+1} is called a graph of the level $k + 1$. In this case the graph $\Gamma_1 = \langle\{V_{\alpha_1}^1 | \alpha_1 \in \Lambda_1\}, \{u_{\alpha_1,\beta_1}^1 | \alpha_1, \beta_1 \in \Lambda_1\}\rangle$ is called a quasi-fractal graph. If all the graphs $\Gamma_k = \langle\{V_{\alpha_k}^k|\alpha_k \in \Lambda_k\}, \{u_{\alpha_k,\beta_k}^k|\alpha_k, \beta_k \in \Lambda_k\}\rangle, r = 1, \ldots, n, \ldots$ are isomorphic to each other, then the graph $\Gamma_1 = \langle\{V_{\alpha_1}^1|\alpha_1 \in \Lambda_1\}, \{u_{\alpha_1,\beta_1}^1|\alpha_1, \beta_1 \in \Lambda_1\}\rangle$ is called a fractal graph. The graph $\Gamma_k = \langle\{V_{\alpha_k}^k|\alpha_k \in \Lambda_k\}, \{u_{\alpha_k,\beta_k}^k|\alpha_k, \beta_k \in \Lambda_k\}\rangle$, where $\{V_{\alpha_k}^k|\alpha_k \in \Lambda_k\}$ is the set of vertices of the graph Γ_k, $\{u_{\alpha_k,\beta_k}^k|\alpha_k, \beta_k \in \Lambda_k\}$ is the set of edges of the graph Γ_k, is called a quasi—fractal graph of the level k.

The notion of a quasi-fractal graph is a generalization of the notion of a graph, since an ordinary graph is a quasi-fractal graph of the first level.

Examples An example of a fractal graph without edges is the Dirichlet function:

$$D(x) = \begin{cases} 1, x \in Q \\ 0, x \in R\backslash Q \end{cases}$$

It is well known that $D(x) = \lim_{m\to\infty} \lim_{n\to\infty} \cos^{2n}(m!\pi x)$. The Dirichlet function is periodic one with a period equal to any rational number:

$$D(x) = D(x + q) \text{ for any } q \in Q.$$

2.8.1 Decomposition and Synthesis Based on Quasi-fractal System Models

Thus, the concept of fractal (respectively quasi-fractal graph) defines, in fact, the structure of the simulated system, and thus the process of decomposition of the original system into subsystems.

A quasi-fractal algebraic system representing the model of the original system S also determines the synthesis of the system. Let's consider some examples.

Examples The test task of a level not exceeding k is represented by the formula of NPC, that is narrow predicate calculus, $F(x_1, x_2, \ldots, x_n)$, where variables $x_1, x_2, \ldots, x_n$ take values in the domains $D_m, m \leq k$, modeled by the graphs $\Gamma_m, m \leq k$, on the set of graphs $\Gamma_m, m \leq k$.

2.8.2 Appendix. Quasi-fractal Systems Recognition Methods. Monitoring Learning Outcomes Through Testing

Monitoring of learning outcomes can be carried out by different methods. But none of them allows one to determine the true level of the knowledge level of academic discipline by students. In this regard, it is desirable that the conditions for monitoring learning outcomes will be the same for all students. One of the methods that satisfies this requirement is the method of pedagogical testing. Herewith the problem of teacher testing is the mismatch (or partial mismatch) between the information model of knowledge representation of the discipline that is implemented in the learning process, on the one hand, and the information model of teaching test (structure and content of test tasks), on the other hand.

Let's consider one of the possible ways to resolve this problem. To provide the compliance of testing purposes, the learning objectives, as on academic discipline as a whole and on its sections, topics, educational issues, methods of training solution it is required that [1]:

- in the basis of training and control of knowledge the same information model of representation of knowledge about an educational discipline (subject area) was used;
- the information model of representation of knowledge about the subject has admitted the possibility of its decomposition, based on the learning objectives, topics, academic disciplines, the training sessions, training issues, excerpts (including the formulation and solution of typical educational tasks, etc.), on information models of knowledge representation of the relevant parts of the subject area;
- information models of representation of knowledge about the relevant parts of the subject area could be synthesized into more general models of representation of knowledge of different levels, up to the model of representation of knowledge about the subject area.

To ensure that all of the above requirements are met, it is convenient to use semantic knowledge networks as information models for representing knowledge about the subject area and its components. In the most general form, the information model of the semantic knowledge network can be represented as a directed graph [14, 15]:

$$G = \langle \{V_\alpha | \alpha \in \Lambda\}, \{u_{\alpha,\beta} | \alpha, \beta \in \Lambda\} \rangle, \tag{2.7}$$

where $\{V_\alpha | \alpha \in \Lambda\}$ is a set of vertices of the graph G, $\{u_{\alpha,\beta} | \alpha, \beta \in \Lambda\}$ is a set of edges of the graph G..

The analysis of graph (1) showed that all its components can be considered as fractals [16, 17]. In this case, a fractal is a geometric shape that can be divided into parts, where each of the parts is a reduced version of the whole. B. Mandelbrot gave a stricter definition of the fractal than the one given above. According to this definition, a fractal is an object whose Hausdorff—Besikovich dimension is larger than its topological dimension [16]. To describe the test tasks, let's use the language of narrow predicate calculus. Let the information model of knowledge representation about the subject area (or about its part) be a quasi-fractal graph Γ_A, the i-th level of which is represented by the graph

$$\Gamma_i = \langle \{V_{\alpha_i}^i | \alpha_i \in \Lambda_i\}, \{u_{i,\beta_i}^i | \alpha_i, \beta_i \in \Lambda_i\} \rangle, i = 1, 2, \ldots \tag{2.8}$$

Then it is necessary that the information model of the pedagogical test corresponds to the model (2.7), i.e. it has the form T_A:

$$T_i = \langle \{ t_{\alpha_i}^i | \alpha_i \in \Lambda_1 \}, \{ r_{\alpha_i, \beta_i}^i | \alpha_i, \beta_i \in \Lambda_i \} \rangle, i = 1, 2, \ldots \tag{2.9}$$

where $\{ t_{\alpha_i}^i | \alpha_i \in \Lambda_i \}$ is the set of vertices of the graph $T_i, i = 1, 2, \ldots;$ $\{ r_{\alpha_i, \beta_i}^i | \alpha_i, \beta_i \in \Lambda_i \}$ is the set of edges of the graph $T_i i = 1, 2, \ldots$.

Each graph $T_i, i \leq k$, represents the basis for the development of test task 1, 2 and subsequent levels. Moreover, each test task can be presented in the database of test tasks in several ways. So, the pedagogical test can be defined as follows.

Definition 2.12 A pedagogical test consisting of test tasks of the level not exceeding k on the set of graphs $\Gamma_i, i \leq k$, is defined by the NPC (i.e. narrow predicate calculus) formula

$$F(x_1, x_2, \ldots, x_n), \tag{2.10}$$

where the variables $x_1, x_2, \ldots, x_n$ take values in the domains $D_i, i \leq k$, modeled by graphs $T_i, i \leq k$; n is the number of test tasks in the pedagogical test. Thus, the concept of a fractal (respectively quasi-fractal) graph determines, in fact, the structure of the simulated system, which defines the process of decomposition of the original system into subsystems. At the same time, existing knowledge assessment systems (i.e. in this way, and feedback control systems in the system) are not sensitive enough (accurate). In [18–20] another kind of the assessment of the knowledge system was proposed, Knowledge Asset Map. One can see that in common the mentioned approach is very close to the fractal approach as we can refine the assessment by adding new factors taking in attention. We propose the corresponding areas of knowledge assessment introduced in [16] to be evaluated using a quantitative assessment—the Hausdorff—Besikovich dimension.

2.9 Example. The Control Algorithm of a Complex System Based on a Quasi-fractal Model

Here we consider one special case. Let the system S be modeled by a group of determining factors, which is a fractal cyclic group $Z_2 = \langle Z_2|+, -, 0 \rangle$.

Then the system S is an innovative [1] one and each of its determining factors is a fractal—innovative one.

Here

e is a neutral factor, a is a factor, determining the system S. In such a system, the sequential composition (repetitive action) of the factor determining the system neutralizes this factor, or even can destroy the system. For example, a violation of

Fig. 2.4 Cayley table of the
fractal group Z_2,
representing the system S

+	e	a
e	e	a
a	a	e

the proportions in the composition of the powder, exceeding the allowable speed of
the system, exceeding the allowable dosage, etc.

2.9.1 Elementary Controlled Systems

The question arises: Can an algorithm to achieve the goal of the system be built?

Definition 2.13 A system S is called elementary controllable if the elementary theory
of its model G_S is solvable.

Examples

1. Semantic networks represented by finite Boolean algebras.
2. Systems modeled by Abelian groups. In particular, the system from point 2.8,
   example, (which is a fractal cyclic group $Z_2 = \langle Z_2|+, -, 0\rangle$) is elementary
   controllable.

V. Martianov obtained the following results [21]. The elementary (universal)
element-subgroup theory of Abelian groups of the signature $\sigma = \langle +, -\rangle$ is insolv-
able (respectively, solvable), where $+$ is a group operation, $\in$ is the relation of
an element belonging to an Abelian group to a subgroup. Recall that the element-
subgroup theory of Abelian groups uses multi-basic algebraic systems of the form
$A_1, A_2; +, \in$, where A_1 is the set of elements of an Abelian group, A_2 is the set
of subgroups of an Abelian group, $+$ is a group operation, $\in$ is the relation of an
element belonging to an Abelian group to a subgroup. In [21], the decidability of the
element-subgroup theory of abelian torsion-free groups of the signature $\sigma = \langle +, \in\rangle$
with universal quantifiers on elements and with arbitrary quantifiers on subgroup
is proved. In addition, there was announced a result on the solvability of the entire
element-subgroup theory of Abelian groups of signature σ with universal quantifiers
on elements and arbitrary quantifiers on subgroups. The technique proposed by A.
Robinson, based on the notion of model completeness and the notion of elementary
extension of the model, was used to prove this statement in [21]. In fact, the concept
of elementary extension is a generalization of the concept of the continuity of Abelian
groups to a class of arbitrary models.

2.10 Conclusion. Connection of Quasi-fractals with Synergetics

According to Podlazov [6], synergetics has become one of the main sources of new views on the structure of the world, the efforts of which were initially aimed at identifying the universal mechanisms of the structure and functioning of systems of various nature [6, 22, 23]. To date, three paradigms have formed in synergetics [6, 24].

The first of these paradigms is the paradigm of self-organization. In systems that are far from the equilibrium position, self-organization processes occur, leading to the selection of a small number of order parameters—leading variables, to which all the others are tuned out from the set describing the system of quantities [6, 22, 25]. The greatest attention in the construction of this paradigm was paid to structures—states arising from the coordinated behavior of a large number of particles. In spatially distributed systems, in the presence of dissipation, self-organization can lead to a loss of stability of the homogeneous equilibrium state. As a result, stationary structures are formed, which I. R. Prigogine proposed to call dissipative [6, 25], or periodic or non-periodic oscillations develop, which, following R. V. Khokhlov, are called autowave processes [6, 26]. A separate class of structures arising in systems with strong positive feedback are processes developing in the mode with aggravation, actively studied in the scientific school of S. P. Kurdyumov [6].

The second paradigm of synergetics is the paradigm of dynamic chaos. This phenomenon is a complex non-periodic behavior observed in deterministic systems (i.e. those where the future is uniquely determined by the past and present and there are no random factors). The main result at this stage was to establish the existence of limits of predictability associated with the existence of a forecast horizon—the end-time dynamic forecast of the system's behavior becomes impossible. Fundamental concepts such as a strange attractor and trajectory scattering were also introduced, universal scenarios of transition from regular to chaotic motion when changing an external parameter were described.

The third paradigm of synergetics—the paradigm of complexity—lies at the junction of the two previous ones. If the first and second paradigms are associated, respectively, with order and chaos, then the third is usually denoted by the phrase "life on the edge of chaos" [6, 27, 28]. Synergetics is also called the science of complexity, since the existence of such universal mechanisms serves as a guarantee of the cognizability of complex systems. To be realizable, the complex must be arranged simply, and simple schemes are few and therefore universal. The complexity of a system is usually understood as its principal irreducibility to the simple sum of its parts. As it follows from this definition, system complexity is inextricably linked with system non-linearity. In fact, the principle of superposition is applicable to a linear system and it can be decomposed into independent components, from the description of which the original system can be easily assembled. This circumstance is due to another name for synergetics—nonlinear dynamics, or—more broadly—nonlinear science.

It turns out that quasi-fractals and synergetics are closely related. In synergetics, from a statistical point of view, catastrophism is a consequence of scale invariance. Therefore, it is peculiar to the phenomena related to the conduct of the first two paradigms of synergetics, exactly to the extent that these phenomena possess scale-invariant properties. The development of catastrophic events is possible only with the coordinated behavior of various parts of the system, i.e. if it has integral properties. Integrity is the third distinctive feature of the phenomena and processes underlying the paradigm of complexity, which largely determines this direction. All signs of being on the edge of chaos are observed at a critical point (bifurcation point), where there is a change in the number or type of equilibrium states of the system. Bifurcations of the system arise at those points in space at which the system exhibits scale invariance. Apparently, these are fixed points of the Brouwer fixed-point theorem. The results of this section are consonant with the results obtained according to the second paradigm of synergetics about the presence of a forecast horizon.

References

1. Serdyukova, N., Serdyukov, V.: Algebraic Formalization of Smart Systems. Theory and Practice, Smart Innovation, Systems and Technologies, SIST, vol. 91, Springer Nature, Switzerland (2018)
2. Gorelov, M.A.: Fractals and Repetitive Games, Computing Center RAS, Moscow, Home page http://www.mtas.ru/
3. Fursov, V.A.: Identification of systems by a small number of observations: textbook/V. Samar State Aerospace University, A. Fursov-Samara (2007)
4. Kravchenko, A.S.: Invariant Measures of Self-Similar Fractals and Metric Properties Of Self-Affine Curves, Novosibirsk (2006). (in Russian)
5. Hutchinson, J.: Fractals and self similarity. Indiana Univ. Mathem. J. **30**, 713–747 (1981)
6. Podlazov, F.: Theory of Self-Organized Criticality—The Science of Complexity. Institute of Applied Mathematics M.V. Keldysh RAS, Home page, https://mipt.ru/students/organiztion/mezhpr/arxiv/mezhpred2/podlazov
7. Danilov, V.I.: Lectures on fixed points. Russian Economic School, Moscow (2006). (in Russian)
8. Bucur, I., Deleanu, A.: Introduction to the Theory of Categories and Functors, A Wiley—Interscience Publication. Wiley, London, New York, Toronto, Sydney, Reprinted (1970)
9. Kurosh, A.G., Livshits, A.H., Shulgeifer, E.G.: Fundamentals of category theory. UMN **15**(96), 3–52 (1960)
10. Artamonov, V.A., Saliy, V.N., Skornyakov, L.A., Shevrin, L.N., Shulgeifer, E.G.: General algebra. In: Skornyakov L.A. (ed.), Main edition of physical and mathematical literature, vol. 2, Moscow (1991). (in Russian)
11. Neumann, H.: Varieties of Groups, Springer-Verlag (1967), ISBN 978-3-642-8860\-0 ISBN 978-3-642-88599-0 (eBook) 001\0.\007/978-3-642-88599-0
12. Kulakov, Y.A., Vorotnikov, V.V., Gumenyuk, I.V.: Analysis of the structure of a telecommunication network by presenting its topology with a prefractal graph, Visnik NTUU "KPI" Informatics, management and accounting of equipment **58**, 68–72 (2013) (in Russian)
13. Semenov A.S.: Managing Large Systems, Special issue 30.1 "Network models in management", Moscow, 91–103
14. Home page http://www.lib.knigi-x.ru/23raznoe/728437-1-upravlenie-bolshimi-sistemami-specialniy-vipusk-301-setevie-modeli-upravlenii-udk-00.php (in Russian)
15. Ore, O.: Graph Theory, Science. Moscow (1980) (in Russian)

16. Harari, F.: Graph theory, Mir, Moscow (1973) (in Russian)
17. Mandelbrot, B.B.: The fractal geometry of nature. Am. J. Phys. **51**(3), 1–468 (1982)
18. Semenov, A.S.: Fractal evolving architecture. In: Novikov, D.A. (ed.). Collection of scientific papers "Management of large systems", 91–103, Institute of Management Problems named after V. A. Trapeznikov of the Russian Academy of Sciences, Moscow (2010) (in Russian)
19. Travica, B.: University of Manitoba, conceptualizing knowledge culture. J Appl Knowl Manag A Publ Int Inst Appl Knowl Manag **1**(2), 85–104 (2013)
20. Boisot, M.: Knowledge Assets: Securing Competitive Advantage in the Information Economy. Oxford University Press, Oxford, UK (1998)
21. Boisot, M.: Information and Organizations: The Manager as Anthropologist. Fontana, London (1987)
22. Martyanov, V.I.: About the element-subgroup theory of Abelian groups, The Bulletin of Irkutsk State University. Series Mathematics **4**(3), 99–109 (2011) (in Russian)
23. Haken, G.: Synergetics, Mir, Moscow (1980) (in Russian)
24. Knyazeva, E.N., Kurdyumov, S.P.: Bases of synergetics. Modes with peaking, self-organization, temporary, Aleteia, Saint Petersburg (2002) (in Russian)
25. Malinetsky, G.G., Potapov, A.B.: Modern problems of nonlinear dynamics, Ed. 2nd, corrected and additional, Editorial URSS, Moscow (2002) (in Russian)
26. Nikolis, G., Prigozhin, I.: Self-Organization in Non-Equilibrium Systems, Mir, Moscow (1979) (in Russian)
27. Zykov, V.S.: Modeling of wave processes in excitable media. Nauka, Moscow (1984). (in Russian)
28. Waldrop, M.M.: Complexity: The Emerging Science at the Edge of Order and Chaos. Touchstone, New York (1993)

Chapter 3
General System Function

Large-scale invariance, or its self-similarity of
fractal structure, is its
characteristic property
F. A. Tzitsin, Fractal Universe,
Home page
http://www.delphis.ru/journal/article/fraktalnaya-vselennaya

Abstract In this chapter the following basic questions are considered:

- natural fractal algebraic systems through series of examples from mathematics, natural sciences, telecommunications,
- the concept of quasi-fractal homomorphism and target quasi-fractal of a system, system target functions,
- regulation of a quasi-fractal system. Regulatory functions of the system. Theorems explaining the appearance of mutations.

Keywords Testing · Binary tests · System efficiency · Knowledge system · Semantic network · Probability measure

3.1 Introduction

We shall approach the description of system functions from the point of view of quasi-fractal algebraic systems introduced in Chap. 2. Before doing this, we define natural fractal (algebraic) systems and show that they can be defined in the form of quasi-fractal algebraic systems. Writing natural fractals in the form of a quasi-fractal algebraic system allows one to represent the process of decomposition of the system when moving from the upper levels of the quasi-fractal algebraic system to the lower ones, and, thus, allows to describe more clearly the functions of the system. According to Surmin [1], the functional organization of a system is that each system is distinguished by its own set of external functions. Their implementation makes its elements and subsystems function in a certain way in the direction of

© Springer Nature Switzerland AG 2021

N. A. Serdyukova and V. I. Serdyukov, *Algebraic Identification*
of Smart Systems, Intelligent Systems Reference Library 191,
https://doi.org/10.1007/978-3-030-54470-6_3

achieving the external functions' aims of the system as a whole. Functional connections arise between the elements due to this. The authors of the book "Principles of the organization of social systems: Theory and practice" determine the effect of the organization's functionality: "In organization theory, a functional relationship is such a form of interaction between the corresponding elements of the whole, in the presence of which the state and behavior of these elements are interdependent, and the chain of causes and effects is closed" [2, p. 46]. Functional organization is clearly manifested in social systems in which the distribution of the following functions is observed: managerial; political; informational; educational; environmental impact; technical; team; economic. The operational functions of the system are associated with the choice of the method of activity, environmental impact. In connection with the functional organization of the system, the following behavior strategies can be applied:—minimax—this is a guide to an unfavorable situation. But the result could not be worse than intended. The minimax strategy guarantees: it can be better; it cannot be worse;

– minimum average risk—an advantage in high efficiency on average. The main disadvantage of this strategy is the difficulty in determining the average risk;
– acceptable risk. A relatively high risk is allowed and a way of behavior is sought in which success will be maximized. The main disadvantage is the difficulty in determining the amount of acceptable risk. The functions of the system are also described in Chap. 5 from [1, p. 132].

3.2 Natural Fractal Algebraic Systems

Let's begin with preliminary remarks and some examples.

1. **Examples in Algebra**. Consider the notion of an algebraic system, and the notion of algebra in the sense of Maltsev [3]. By isolating subsystems of an algebraic system, subalgebras of an algebra, and then subsystems of subsystems of algebraic systems, subalgebras of subalgebras of algebras, and continuing this process by induction, we obtain a fractal structure on algebraic systems or on algebras.
2. **Examples in medicine, in the natural sciences, in telecommunications**. Currently, fractals are used in medicine, in the natural sciences, in telecommunications. There exists a theory of the infinite nesting of matter [4, 5]. In the book of known member of Russian academy of cosmonautics Basil D. Shabetnik Fractal Physics Science of the Universe [5], for the first time in history a new science about nature, human being and consciousness is represented on the basis of unite foundation of the Universe—electric charge. Fractal Physics is a knowledge of the Universe, of the world essence as the traditional physics studies separate its parts. In Fractal Physics a conformity to natural laws is discovered with the aid of which the order and harmony of the Universe is achieved. The notion of unite nature electromagnetic comes to discovery a law of unite interaction. A geometry and structure of material objects lead to distinctive electromagnetic

effects manifested in different forms and known us as gravitation, strong and weak interaction and electromagnetic one. The theory of the Infinite nesting of matter (fractal theory) as opposed to atomism is a scientific theory based on inductive logical conclusions about the structure of the observable Universe and emphasizing the hierarchical organization of nature: from the smallest observed elementary particles to the largest visible clusters of galaxies. In this book the fact that the global hierarchy of nature is discrete is highlighted; the atomic, stellar, and galactic levels are especially distinguished. The fact that the global hierarchy of nature is discrete is highlighted in Fractal Physics. It states that for each class of objects or phenomena at a given scale level, there is a similar class of objects or phenomena at any other scale level. Self-similar analogues of objects and phenomena from different levels have the same morphology, kinematics and dynamics. Thus, the theory claims that any particle has its own system of particles, and an electromagnetic wave consists of electromagnetic waves. Considering the structure of the atom at the micro level and the classification of elementary particles in physics, one also comes to a fractal structure. In [6], features of the proton fractal are highlighted. It is shown that:

1. The proton fractal is not a closed structure. Such constructions are typical for objects that interact with the external environment.
2. The proton fractal forms a unique tree structure. The fractal tree in it is formed by the alternation of symmetric and asymmetric branches. This means that the quantitative characteristics of fractal branches for symmetric branches are equal, and for asymmetric branches they are sharply different, and this difference increases with the growth of the fractal structure.
3. Both in the structure of the proton fractal and in the fractal formula of the proton, binarity is clearly manifested. From the point of view of the structure this means that each current substructure is formed by two lower order substructures.
4. The proton fractal is a deterministic fractal and reflects the deterministic process of the structural genesis of the substance.
5. The proton fractal is a "folding", "converging" fractal, unlike most tree-shaped fractals, which are characterized by a "unfolding" structure.
6. The proton fractal reflects not so much a fixed cluster structure as the dynamics of the proton structure formation process.
7. The proton fractal is not an infinite structure, but has a fixed scale, in which the structure is completed at the 20th step of the structure genesis of substance.

Definition 3.1 A quasi-fractal is called natural if each of its elements is a subsystem of the original system under study.

So, characteristic property of a natural fractal is the following one:
In natural fractals, each element is a subsystem of the original system under study. One gets from here that fractals from Examples 3.1 and 3.2 natural ones.
Let's show that for Example 3.1 show that in the first case one can write a natural algebraic fractal in the form of a quasi-fractal algebraic system. Recall the essence

of definition of a quasi-fractal algebraic system from Chap. 2 in order to highlight some important details:

Let's consider an algebraic system $A_1 = \langle A_1; \Omega_1 \rangle$ of the signature Ω_1, and what is more let every element $a_\alpha, \alpha \in \Lambda_1$, of the main set A_1 of a system A_1 in turn be an algebraic system of the signature Ω_2. That is:

$a_\alpha = A_2^\alpha = \langle A_2^\alpha; \Omega_2 \rangle$, $\alpha \in \Lambda_1$ is an algebraic system of the second fractal level. Let's continue this process by induction. If an algebraic system $A_k^\alpha = \langle A_k^\alpha; \Omega_k \rangle$, $\alpha \in \Lambda_k$ is an algebraic system of the fractal level k, and every element $a_\alpha, \alpha \in \Lambda_k$ of the main set A_k^α of a system A_k^α is an algebraic system $a_\alpha = A_{k+1}^\alpha = \langle A_{k+1}^\alpha; \Omega_{k+1} \rangle$ of the signature Ω_{k+1}, of the fractal level $k + 1$, $\alpha \in \Lambda_{k+1}$, then the algebraic system $A_1 = \langle A_1; \Omega_1 \rangle$ is called a quasi-fractal algebraic system.

We can distinguish two different cases:

Now consider a special case when the number of subsystems of each fractal of the (considered) algebraic system included in the quasi-algebraic system are less than the number of elements of the basic set of the original algebraic system $A_1 = \langle A_1; \Omega_1 \rangle$. Let elements of the level 2 be subsystems of a system A_1, elements of the level 3 be subsystems of subsystems of a system A_1, and so on, then the process continues by induction, and so on.

Let us now consider the general case when we do not know what is more: the number of elements of the basic set of the algebraic system under consideration or the number of its subsystems. In the most general case, for example, if we are talking about subalgebras of the algebra of sets, $A_1 = \langle A_1; \Omega_1 \rangle$ is the algebra of sets, the number of its subalgebras is equal to $2^{|A_1|}$, where $|A_1|$ is the number of elements of the main set of algebra of sets A_1. Let's define linear order relations $\rho_k, k = 1, 2, \ldots.$[1] on the sets Λ_k. After that we define that at the second level of the quasi-fractal, each element $a_\alpha, \alpha \in \Lambda_2$ will have $2^{|A_1|}$ elements, each of which is a subsystem of the original algebraic system $A_1 = \langle A_1; \Omega_1 \rangle$. Then this process continues by induction. At the step with number k each element of the set A_k^α will be a set of $2^{2^{\cdot^{\cdot^{2^{|A_k^\alpha|}}}}}$ elements, and each element of this set will be a subsystem of an algebraic system $A_k^\alpha = \langle A_k^\alpha; \Omega_k \rangle$.

Remark Note that, in fact, the representation of a natural fractal in the form of a quasi-fractal algebraic system gives an idea of the decomposition process of the system during the transition from the upper levels of the quasi-fractal to the lower.

General system's functions

We shall model system's functions using group of homomorphisms of the group G_S factors determining the system S into $G_{S'}$ where S' is the desired state of a system S (optimal state of system S, for example).

So, we have a homomorphism

---

[1] A total partially ordered set is a partially ordered set that has a least element that is the only element that precedes any other element and each directed subset of which has an exact upper bound [7, 8]. From the axiom of choice follows the Principle of total ordering (Zermelo's theorem). The principle of total ordering is in that any set can be done total ordered [9].

$$f : G_S \rightarrow G_{S'}$$

Then by the theorem about homomorphisms (first isomorphism theorem [3]) we have:

$$Im f \cong G_S / Ker f$$

The question arises:

What is it $Ker f$, or rather, how to interpret $Ker f$?

In the economy example [10], $Ker f$ is the shadow sector, the inefficient sector, the undesirable sector, etc.

Representation of system's functions in the form of homomorphisms of a group of factors that determine the system allows one to obtain the following classification:

(1) If all the properties get into the kernel $Ker f$, then the system cannot be fully identified, that is, when you turn off any of its subsystems, it already loses its unique set of properties.

(2) What are in the kernel $Ker f$, do not differ. The system is identified by the set of properties contained in the kernel $Ker f$: if one disables any property from $Ker f$ the system loses the entire unique set of properties from $Ker f$.

3.3 The Concept of Quasi-fractal Homomorphism. Target Quasi-fractal of a System. System Target Functions

Definition 3.2 Let $A_1 = \langle A_1; \Omega_1 \rangle$, $B_1 = \langle B_1; \Omega_1 \rangle$ be quasi-fractal algebraic systems such that for each level of a quasi-fractal systems the systems $A_k^\alpha = \langle A_k^\alpha; \Omega_k \rangle$ and $B_k^\alpha = \langle B_k^\alpha; \Omega_k \rangle$ have one the same signature Ω_k. A mapping $f : A_1 = \langle A_1; \Omega_1 \rangle \rightarrow B_1 = \langle B_1; \Omega_1 \rangle$ is called a quasi–fractal homomorphism from a system $A_1 = \langle A_1; \Omega_1 \rangle$ into a system $B_1 = \langle B_1; \Omega_1 \rangle$, if the following two conditions are fulfilled:

(1) For every $\alpha \in \Lambda_k$ the equality $f(a_\alpha) = f(b_\alpha)$ holds for some $b_\alpha \in \Lambda_k$, that is f maps the elements of the quasi-fractal algebraic system $A_1 = \langle A_1; \Omega_1 \rangle$ of the level $k + 1$ into elements of the same level $k + 1$ of the quasi-fractal algebraic system $B_1 = \langle B_1; \Omega_1 \rangle$;

(2) f saves all operations and predicates of the signature Ω_k for all levels k of the quasi-fractal. That is:
$f\left(F_\xi(x_1, \ldots, x_{m_\xi})\right) = \left(f\left(F_\xi\right)\right)\left((f(x_1), \ldots, f(x_{m_\xi})\right)$ for every operation $F_\xi \in \Omega_k$ and every $x_1, \ldots, x_{m_\xi} \in A_k^\alpha$
and
$P_\eta\left((x_1, \ldots, x_{n_\eta})\right) \Leftrightarrow \left(f(P_\eta)\right)\left((f(x_1), \ldots, f(x_{n_\eta})\right)$ for every predicate $P_\eta \in \Omega_k$ and every $x_1, \ldots, x_{n_\eta} \in A_k^\alpha$

Let's recall Definition 2.1 from Chap. 2 for convenience.

Definition 2.1 Let's consider the algebraic system $A_1 = \langle A_1; \Omega_1 \rangle$ of the signature Ω_1, such that every element a_α, $\alpha \in \Lambda_1$, of the main set A_1 of the system A_1 in turn is an algebraic system of the signature Ω_2. That is $a_\alpha = A_2^\alpha = \langle A_2^\alpha; \Omega_2 \rangle$, $\alpha \in \Lambda_1$ is an algebraic system of the second level. Continue this process by induction. If an algebraic system

$A_k^\alpha = \langle A_k^\alpha; \Omega_k \rangle$, $\alpha \in \Lambda_k$ is an algebraic system of the level k of the fractal and every element a_α, $\alpha \in \Lambda_k$ of the main set A_k^α of the system A_k^α is an algebraic system $a_\alpha = A_{k+1}^\alpha = \langle A_{k+1}^\alpha; \Omega_{k+1} \rangle$ of the signature Ω_{k+1}, of the level $k+1$, $\alpha \in \Lambda_{k+1}$, of the fractal, then the algebraic system $A_1 = \langle A_1; \Omega_1 \rangle$ is called a quasi-fractal algebraic system. If all signatures Ω_k, $k = 1, \ldots n, \ldots$, are equal to each other and all the systems $\langle A_k^\alpha; \Omega_k \rangle$, $\alpha \in \Lambda_k$, are isomorphic to each other then the algebraic system $A_1 = A_1; \Omega_1$ is called a fractal algebraic of the signature Ω_1.

Remark Let's note that, in fact, a quasi-fractal algebraic system allows us to represent the processes that occur in the system under the influence of impacts (represented by functions) in the form of discrete (by quasi-fractal levels) system states obtained as a result of the influences.

Let now $G_S = \langle G_S, *, ^{-1} \rangle = A_1 = \langle A_1; \Omega_1 \rangle$ be a quasi-fractal group of determining factors modelling the system S, that is at each level k of a quasi-fractal $G_S = \langle G_S, *, ^{-1} \rangle = A_1 = \langle A_1; \Omega_1 \rangle$ an algebraic system $A_{k+1}^\alpha = \langle A_{k+1}^\alpha; \Omega_{k+1} \rangle$ is a group, that is $A_{k+1}^\alpha = \langle A_{k+1}^\alpha; \Omega_{k+1} \rangle = G_{k+1}^\alpha = \langle G_{k+1}^\alpha; *, ^{-1} \rangle$ are groups.

We should explain that in this notation $A_1 = \langle A_1; \Omega_1 \rangle$ plays the role of a universal variable to denote a quasi-fractal algebraic system.

Let's consider a quasi-fractal algebraic system $Hom_1(G_S, G_S) = B_1 = \langle B_1; \Omega_1 \rangle$ such that at each quasi-fractal level k of the algebraic quasi-fractal system $B_1 = \langle B_1; \Omega_1 \rangle$ the algebraic system $B_{k+1}^\alpha = \langle B_{k+1}^\alpha; \Omega_{k+1} \rangle$ is a semigroup of homomorphisms $Hom(G_{k+1}^\alpha, G_{k+1}^\alpha) = < Hom(\langle G_{k+1}^\alpha; *, ^{-1} \rangle, \langle G_{k+1}^\alpha; *, ^{-1} \rangle, *, ^{-1}) >$, with respect to the operation of composition of quasi-homomorphisms. If a group G_S is an abelian group then, the operation $*$ is the operation of addition of homomorphisms.

Definition 3.3 A quasi-fractal algebraic system $Hom_1(G_S, G_S)$ is called a quasi-fractal semigroup of homomorphisms of a quasi-fractal group $G_S = \langle G_S, *, ^{-1} \rangle$ of the determining factors which models the system S.

3.3.1 Algorithm of Construction a Level of a System Predictability

1. Next, according to the algorithm described in Chap. 2, Examples 1, 2, 3, we construct a compressive map of a quasi-fractal algebraic system $Hom_1(G_S, G_S)$. After that we define the metric d on the quasi-fractal $Hom_1(G_S, G_S)$ in accordance with Example 4, Chap. 2.

2. We embed the metric quasi-fractal $\boldsymbol{Hom}_1(\boldsymbol{G_S}, \boldsymbol{G_S})$ in the completion by this
    metric, which is a metric compact with metric d. According to [11], this complete
    metric compactum, when its dimension equals to n, is homeomorphic to the unit
    ball D_n in R^n, where $D_n = \{x \in R^n \| |x| \leq 1\}$ and $|x| = |x_1, \ldots, x_n| =$
    $\left(x_1^2, \ldots, x_n^2\right)^{\frac{1}{2}}$.
3. Then, according to Brouwer's theorem, we obtain a fixed point of the compres-
    sive map from Example 4, Chap. 2, which defines the predictability level of
    the system S. In our case, the fixed point of the compressing map is a group
    of homomorphisms $Hom\left(G_{k+1}^\alpha, G_{k+1}^\alpha\right)$ of a group G_{k+1}^α of some group of the
    level k of a quasi-fractal $\boldsymbol{G_S} = \langle G_S, *, ^{-1}\rangle = A_1 = \langle A_1; \Omega_1\rangle$. We shall
    call $Hom\left(G_{k+1}^\alpha, G_{k+1}^\alpha\right)$ a fixed or invariant semigroup of homomorphisms of
    a quasi-fractal $\boldsymbol{Hom}_1(\boldsymbol{G_S}, \boldsymbol{G_S})$.

Remark The fixed semigroup of homomorphisms $Hom\left(G_{k+1}^\alpha, G_{k+1}^\alpha\right)$ of a quasi-
fractal $\boldsymbol{Hom}_1(\boldsymbol{G_S}, \boldsymbol{G_S})$ defines the target functions of the system S.

3.4 Regulating Quasi-fractal of a System. Regulatory Functions of the System. Some Theorems Explaining the Appearance of Mutations

The regulation function of the system S is the reaction function of the system S onto
the environmental impact on the functioning of the system S, determined by its target
functions.

Let's define the quasi-fractal system $\boldsymbol{Ker}_1\boldsymbol{f} = \langle Ker_1 f, *\rangle = A_1 = \langle A_1; \Omega_1\rangle$, $f \in \boldsymbol{Hom}_1(\boldsymbol{G_S}, \boldsymbol{G_S})$ as follows

Definition 3.4 Let's choose on the every level k of a quasi-fractal
$\boldsymbol{Hom}_1(\boldsymbol{G_S}, \boldsymbol{G_S}) = A_1 = A_1; \Omega_1$ the set $R_1(\boldsymbol{G_S}, \boldsymbol{G_S})$ of homomorphisms $R_{k+1}^\alpha = \{f : G_{k+1}^\alpha \to G_{k+1}^\alpha | Ker f \neq \{e\}\} \subseteq Hom\left(G_{k+1}^\alpha, G_{k+1}^\alpha\right)$ such that $Ker f \neq \{e\}$
for any $f \in R_{k+1}^\alpha$, any k, any α. Then $\boldsymbol{R}_1(\boldsymbol{G_S}, \boldsymbol{G_S}) = A_1 = \langle A_1; \Omega_1\rangle$ is a
quasi-fractal without algebraic operations on it. Let's define a metric d on the quasi-
fractal $\boldsymbol{R}_1(\boldsymbol{G_S}, \boldsymbol{G_S})$ in accordance with Example 2, Chap. 2. We embed the metric
quasi-fractal $\boldsymbol{R}_1(\boldsymbol{G_S}, \boldsymbol{G_S})$ into the completion by this metric, which is a metric
compact with metric d. According to [11], this is a complete metric compact, in
the case when its dimension is n, it is homeomorphic to the unit ball D_n in R^n,
where $D_n = \{x \in R^n \| |x| \leq 1\}$ and $|x| = |x_1, \ldots, x_n| = \left(x_1^2, \ldots, x_n^2\right)^{\frac{1}{2}}$. Then,
according to Brouwer's theorem, we obtain a fixed point of the contraction map
from Example 4, Chap. 2, which defines the predictability level of the system S. In
our case, the fixed point of the contraction map is the set of homomorphisms $R_{k+1}^\alpha = \{f : G_{k+1}^\alpha \to G_{k+1}^\alpha | Ker f \neq \{e\}\} \subseteq Hom\left(G_{k+1}^\alpha, G_{k+1}^\alpha\right)$ of a group группы G_{k+1}^α
of some level k of a quasi-fractal $\boldsymbol{G_S} = \langle G_S, *, ^{-1}\rangle = A_1 = \langle A_1; \Omega_1\rangle$. We shall
call R_{k+1}^α a fixed or unchanging or invariant set of regulatory homomorphisms of the
quasi-fractal $\boldsymbol{R}_1(\boldsymbol{G_S}, \boldsymbol{G_S}) = A_1 = \langle A_1; \Omega_1\rangle$.

3.4.1   Geometric Examples of Quasi-fractal Groups

We shall use in the examples [12, p. 63]:

the representation of a group as a network consisting of directed segments (edges) whose vertices (or sometimes edges) correspond to elements of the group, and edges correspond to multiplication by the forming elements of the group and their inverse elements, introduced by Cayley in the nineteenth century, and then used by Van Kampen. Rotations of a triangle in a plane form a third-order cyclic group, see Example 3.1 (Fig. 3.1).

The rotations of a square in the plane of the same square form a cyclic group of the fourth order, see Example 3.2 (Fig. 3.2).

The rotation of a regular hexagon in the plane of the same hexagon form a sixth-order cyclic group, see Example 3.3 (Fig. 3.3).

Example 3.1 A system modeled by a third-order quasi-fractal cyclic group $Z_3 = \langle Z_3, *, \ ^{-1} \rangle = A_1 = \langle A_1; \Omega_1 \rangle$

Let's consider the system S, modeled by a quasi-fractal groups of determining factors $G_S = \langle G_S, *, \ ^{-1} \rangle = A_1 = \langle A_1; \Omega_1 \rangle$, which represent the system. Let at each level k of a quasi-fractal $A_1 = \langle A_1; \Omega_1 \rangle$ a group G_{k+1}^α is a cyclic group of the third order. Then a quasi-fractal group of factors

Fig. 3.1   Example 1

Fig. 3.2   Example 2

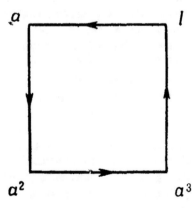

Fig. 3.3 Example 3

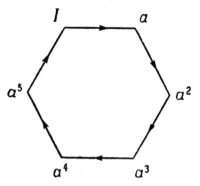

$G_S = \langle G_S, *, \ ^{-1} \rangle = A_1 = \langle A_1; \Omega_1 \rangle$ is called a quasi-fractal group cyclic group of the third order. It can be represented with the help of the following fractal—as a group of rotations of a regular triangle at each fractal level [12, p. 63] (Fig. 3.4).

It is known [14] that $Hom(Z_m, C) \cong C[m]$, so $Hom(Z_3, Z_3) \cong Z_3$. In accordance with it the system S, which is modeled by a quasi-fractal cyclic group of factors $Z_3 = \langle Z_3, *, \ ^{-1} \rangle = A_1 = \langle A_1; \Omega_1 \rangle$ has three target functions f_0, f_1, f_2, defined as follows. Let every level of a quasi-fractal cyclic group $Z_3 = \langle Z_3, *, \ ^{-1} \rangle = A_1 = \langle A_1; \Omega_1 \rangle$ be represented with the help of Cayley table (Fig. 3.5).

Then the target functions of the system S at each level of the quasi-fractal $Hom_1 = (Z_3, Z_3) = B_1 = \langle B_1; \Omega_1 \rangle$ are defined as follows:

$f_0(e) = e, f_0(a_1) = e, f_0(a_2) = e \ f_1(e) = e, f_1(a_1) = a_1, f_1(a_2) = a_2,$
$f_1(e) = e, f_1(a_1) = a_2 f_1(a_2) = a_1.$

This can be interpreted as follows:

Fig. 3.4 Barnsley fern obtained by the extreme modification of the Sierpinski triangle [13]

Fig. 3.5 Cayley table of a
group Z_3

*	e	a_1	a_2
e	e	a_1	a_2
a_1	a_1	a_2	e
a_2	a_2	e	a_1

f_0—the target of the system is a state of rest (all factors ceased to function),
f_1—the target of the system is the functioning of factors a_1 and a_2 without their interchangeability.

f_2—the target of the system is to replace the factor a_1 by a_2 and a_2 by a_1.

The regulatory function of the system S, modeled by a third-order quasi-fractal cyclic group is the function f_0, since only this function has a nonzero kernel. Thus, in this case, the regulation consists in stopping the functioning of the system S.

Similar statements hold for any system modeled by a quasi-fractal cyclic group of factors Z_p of a prime order p, that determine the system S.

Example 3.2 A system modeled by a quasi-fractal cyclic group $Z_4 = \langle Z_4, *, \ ^{-1} \rangle = A_1 = \langle A_1; \Omega_1 \rangle$ of the fourth order

It can be represented using the following fractal (as a group of square rotations at each fractal level) (Fig. 3.6):

Let's consider the system S, which is modeled by a quasi-fractal groups of factors $G_S = \langle G_S, *, \ ^{-1} \rangle = A_1 = \langle A_1; \Omega_1 \rangle$, which represent the system. Let at each level k of a quasi-fractal $A_1 = \langle A_1; \Omega_1 \rangle$ a group G_{k+1}^α is a cyclic group of the fourth order. Then the quasi-fractal groups of factors $G_S = \langle G_S, *, ^{-1} \rangle = A_1 = \langle A_1; \Omega_1 \rangle$ is called a quasi-fractal cyclic group of the forth order and it will be designated

Fig. 3.6 Ref. [13]

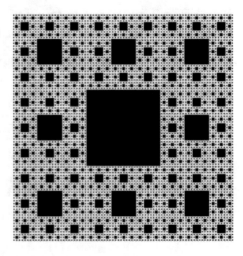

Fig. 3.7 Cayley table of a group Z_4

*	e	a_1	a_2	a_3
e	e	a_1	a_2	a_3
a_1	a_1	a_2	a_3	e
a_2	a_2	a_3	e	a_1
a_3	a_3	e	a_1	a_2

as $Z_4 = \langle Z_4, *, ^{-1} \rangle = A_1 = \langle A_1; \Omega_1 \rangle$. Each level of a quasi-fractal group $Z_4 = \langle Z_4, *, ^{-1} \rangle = A_1 = \langle A_1; \Omega_1 \rangle$ can be represented with the help of Cayley table (Fig. 3.7).

It is known [14] that $Hom(Z_m, C) \cong C[m]$, so $Hom(Z_4, Z_4) \cong Z_4$, Then the target functions of the system S at each level of the quasi-fractal $Hom_1(Z_4, Z_4) = B_1 = \langle B_1; \Omega_1 \rangle$ are defined as follows:

$f_0(e) = e$, $f_0(a_1) = e$, $f_0(a_2) = e$, $f_0(a_3) = e$, is a zero homomorphism, a kernel $Ker f_0$ contains all the factors which define S,

$f_1(e) = e$, $f_1(a_1) = a_1$, $f_1(a_2) = a_2$, $f_1(a_3) = a_3$ Is the identity homomorphism, a kernel $Ker f_1$ is zero,

$f_2(e) = e$, $f_2(a_1) = e$, $f_2(a_2) = a_2$, $f_2(a_3) = e$, a kernel $Ker f_2$ contains e and a_2,

$f_3(e) = e$, $f_3(a_1) = a_3$, $f_3(a_2) = a_2$, $f_3(a_3) = a_1$, a kernel $Ker f_3$ is zero. Hence from one has:

f_0 the target of the system is a state of rest (all factors ceased to function),
f_1 the target of the system is the functioning of all factors without their interchange-ability,
f_2 the target of the system is to stop the functioning of factors a_1 и a_3,
f_3 the target of the system is the interchangeability of factors a_1 and a_3

The regulatory functions of the system S, modeled by a fourth-order quasi-fractal cyclic group, are the function f_0, which stops the functioning of all factors, and the function f_2, which stops the functioning of factor a_2.

Example 3.3 A system modeled by a quasi-fractal cyclic group of the sixth order $Z_6 = \langle Z_6, *, ^{-1} \rangle = A_1 = \langle A_1; \Omega_1 \rangle$

It can be represented using the following fractal (as a group of regular hexagon's rotations at each fractal level) (Fig. 3.8):

Let each level of a quasi-fractal cyclic group of the sixth order $Z_6 = \langle Z_6, *, ^{-1} \rangle = A_1 = \langle A_1; \Omega_1 \rangle$ is represented by Cayley table (Fig. 3.9).

It is known [14] that $Hom(Z_m, C) \cong C[m]$, so $Hom(Z_6, Z_6) \cong Z_6$. Then the target functions of the system S at each level of the quasi-fractal $Hom_1(Z_6, Z_6) = B_1 = \langle B_1; \Omega_1 \rangle$ are defined as follows:

Fig. 3.8 Hex "fractal carpet" [15]

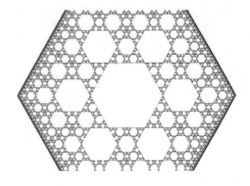

Fig. 3.9 Cayley table of a cyclic group Z_6 of the sixth order

*	e	a_1	a_2	a_3	a_4	a_5
e	e	a_1	a_2	a_3	a_4	a_5
a_1	a_1	a_2	a_3	a_4	a_5	e
a_2	a_2	a_3	a_4	a_5	e	a_1
a_3	a_3	a_4	a_5	e	a_1	a_2
a_4	a_4	a_5	e	a_1	a_2	a_3
a_5	a_5	e	a_1	a_2	a_3	a_4

$f_0(e) = e$, $f_0(a_1) = e$, $f_0(a_2) = e$, $f_0(a_3) = e$, $f_0(a_4) = e$, $f_0(a_5) = e$, is a zero homomorphism, a kernel $Ker f_0$ contains all the factors which define S,

$f_1(e) = e$, $f_1(a_1) = a_1$, $f_1(a_2) = a_2$, $f_1(a_3) = a_3$, $f_0(a_4) = a_4$, $f_0(a_5) = a_5$ the identity homomorphism, a kernel $Ker f_1$ is zero,

$f_2(e) = e$, $f_2(a_1) = a_2$, $f_2(a_2) = a_4$, $f_2(a_3) = e$, $f_2(a_4) = a_4$, $f_2(a_5) = e$, kernel $Ker f_2$ consists of elements e, a_3, a_5.

$f_3(e) = e$, $f_3(a_1) = a_3$, $f_3(a_2) = e$, $f_3(a_3) = a_3$, $f_3(a_4) = e$, $f_3(a_5) = a_3$, kernel $Ker f_3$ consists of elements e, a_2, a_4.

$f_4(e) = e$, $f_4(a_1) = a_4$, $f_4(a_2) = a_2$, $f_4(a_3) = e$, $f_4(a_4) = a_4$, $f_4(a_5) = a_2$, kernel $Ker f_4$ consists of elements e, a_3

$f_5(e) = e$, $f_5(a_1) = a_5$, $f_5(a_2) = a_1$, $f_5(a_3) = e$, $f_5(a_4) = a_4$, $f_5(a_5) = a_3$, kernel $Ker f_4$ consists of e, a_3.

Hence from one has:

f_0 the target of the system is a state of rest (all factors ceased to function),

f_1 the target of the system is the functioning of all factors without their interchangeability, the possibilities for changing the properties of the system are exhausted,

f_2 the target of the system is to stop the functioning of factors a_3 and a_5, and to replace factor a_2 by a factor a_4,

f_3 the target of the system is the target of the system is to stop the functioning of factors a_2 and a_4 and to replace factor a_1 by a factor a_3, a factor a_5 by a factor a_3.

f_4 the target of the system is to stop the functioning of a factor a_3 and to replace a factor a_5 by a factor a_2, a factor a_1 by a factor a_4.

f_5 the target of the system is to stop the functioning of factor a_3, and to replace a factor a_1 by a factor a_5, a factor a_2 by a factor a_1, a factor a_5 by a factor a_3

The regulatory functions of the system S, modeled by a sixth-order quasi-fractal cyclic group, are the functions:

f_0 which stops the functioning of all factors,

f_2 which replaces factor a_2 by a factor a_4,

f_3 which stops the functioning of factors a_2 and a_4 and replaces a factor a_1 by a factor a_3, a factor a_5 by a factor a_3,

f_4 which stops the functioning of a factor a_3 and replaces a factor a_5 by a factor a_2, a factor a_1 by a factor a_4.

f_5 which stops the functioning of a factor a_3, and replaces a factor a_1 by a factor a_5, a factor a_2 by a factor a_1, a factor a_5 by a factor a_3

Example 3.4 A system modeled by a quasi-fractal group of permutations of the third degree $S_3 = \langle S_3, *, \ ^{-1} \rangle = A_1 = \langle A_1; \Omega_1 \rangle$.

A permutation group of the third degree can be represented as a subgroup of the self-displacements group of a regular tetrahedron [12]:

At each level of a quasi-fractal, it can be represented using a group of self-displacements of a regular triangle: three axial symmetries relative to the axes passing through the vertex and center of the triangle, and three rotations relative to the center of the triangle (Fig. 3.10).

Let every level of a quasi-fractal group $S_3 = \langle S_3, *, \ ^{-1} \rangle = A_1 = \langle A_1; \Omega_1 \rangle$ is

represented by a Cayley table, where $e = \begin{pmatrix} 1\ 2\ 3 \\ 1\ 2\ 3 \end{pmatrix}, a_1 = \begin{pmatrix} 1\ 2\ 3 \\ 1\ 3\ 2 \end{pmatrix},$

Fig. 3.10 Self-displacements of a regular triangle [12]

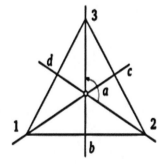

$$a_2 = \begin{pmatrix} 1\ 2\ 3 \\ 3\ 2\ 1 \end{pmatrix}, a_3 = \begin{pmatrix} 1\ 2\ 3 \\ 2\ 1\ 3 \end{pmatrix}, \ a_4 = \begin{pmatrix} 1\ 2\ 3 \\ 2\ 3\ 1 \end{pmatrix}, a_5 = \begin{pmatrix} 1\ 2\ 3 \\ 3\ 1\ 2 \end{pmatrix}.$$

See Fig. 3.11.

The rotation group of a regular triangle in the plane representing subgroup A_3 of even substitutions of a group S_3, is a normal subgroup of a group S_3. It is also a commutant of the group S_3.

It follows from here that the target functions of the third-degree quasi-fractal permutation group $S_3 = \langle S_3, *, \ ^{-1} \rangle = A_1 = \langle A_1; \Omega_1 \rangle$ are the following functions that at each level of the quasi-fractal $Hom_1(S_3, S_3) = B_1 = \langle B_1; \Omega_1 \rangle$ are defined as follows:

$f_0(e) = e$, $f_0(a_1) = e$, $f_0(a_2) = e, f_0(a_3) = e$, $f_0(a_4) = e$, $f_0(a_5) = e$ is a zero homomorphism, the kernel $Ker f_0$ contains all the factors which determined S,

$f_1(e) = e$, $f_1(a_1) = a_1$, $f_1(a_2) = a_2$, $f_1(a_3) = a_3$, $f_0(a_4) = a_4$, $f_0(a_5) = a_5$ is the identity homomorphism, a kernel $Ker f_1$ is zero,

five more homomorphisms in which one symmetry passes into another:

$f_2(e) = e$, $f_2(a_1) = a_2$, $f_2(a_2) = a_1$, $f_3(a_3) = a_3$, $f_2(a_4) = a_4$, $f_2(a_5) = a_5$, the kernel $Ker f_2$ consists of e.

$f_3(e) = e$, $f_3(a_1) = a_2$, $f_3(a_2) = a_1$, $f_6(a_3) = a_4$, $f_3(a_4) = a_5$, $f_3(a_5) = a_3$, $Ker f_3$ co consists of e.

$f_4(e) = e$, $f_4(a_1) = a_3$, $f_4(a_2) = a_2$, $f_4(a_3) = a_1$, $f_4(a_4) = a_4$, $f_4(a_5) = a_5$, $Ker f_4$ consists of e.

$f_4(e) = e$, $f_4(a_1) = a_3$, $f_4(a_2) = a_2$, $f_4(a_3) = a_4$, $f_4(a_4) = a_5$, $f_4(a_5) = a_3$, the kernel $Ker f_4$ consists of e.

$f_5(e) = e$, $f_5(a_1) = a_1$, $f_5(a_2) = a_3$, $f_5(a_3) = a_2$, $f_5(a_4) = a_4$, $f_5(a_5) = a_2$, the kernel $Ker f_5$ consists of e.

$f_6(e) = e$, $f_6(a_1) = a_1$, $f_6(a_2) = a_3$, $f_6(a_3) = a_2$, $f_6(a_4) = a_5$, $f_6(a_5) = a_2$, the kernel $Ker f_6$ consists of e.

Homomorphisms from S_3 into S_3 with nonzero kernels (of two elements) corresponding to epimorphisms:

Fig. 3.11 Cayley table of a group S_3

*	e	a_1	a_2	a_3	a_4	a_5
e	e	a_1	a_2	a_3	a_4	a_5
a_1	a_1	e	a_5	a_4	a_3	a_2
a_2	a_2	a_4	e	a_5	a_1	a_3
a_3	a_3	a_5	a_4	e	a_2	a_1
a_4	a_4	a_2	a_3	a_1	a_5	e
a_5	a_5	a_3	a_1	a_2	e	a_4

$$f_7 \quad \left(\text{to epimorphism } S_3/A_3 \to, \left\langle \{e, a_1\}, *, \quad^{(-1)}\right\rangle\right)$$
$$f_8 \quad \left(\text{to epimorphism } S_3/A_3 \to, \left\langle \{e, a_2\}, *, \quad^{(-1)}\right\rangle\right)$$
$$f_9 \quad \left(\text{to epimorphism } S_3/A_3 \to, \left\langle \{e, a_3\}, *, \quad^{(-1)}\right\rangle\right).$$

It is well known that the group S_3 is perfect, that is, the group $Aut(S_3)$ of automorphisms of the group S_3 is isomorphic to S_3:

$Aut(S_3) \cong S_3$, automorphisms of the group S_3 are the following ones: f_1, f_2, f_3, f_4, f_5

The regulatory functions of the system S, modeled by a quasi-fractal group of permutations of the third degree, are the functions: f_0, f_7, f_8.

In this case, the function f_0 stops the functioning of all factors.

Functions f_7, f_8 stop the functioning of factors a_3, a_4, a_5.

Theorem 3.5 Let the system S be modeled by a quasi-fractal group of permutations of the third degree $S_3 = \left(S_3, *, \quad^{-1}\right) = A_1 = \langle A_1; \Omega_1 \rangle$. At any level of the quasi-fractal representing this system it is not possible to regulate exactly:

- one factor representing the system,
- two factors representing the system,
- four factors representing the system.

When regulating precisely:

- one factor representing the system,
- two factors representing the system,
- four factors representing the system,

system S will change its structure.

So, Theorem 3.5 explains the changes in the structure of the studied system, so one can say that it explains the appearance of mutations.

Example 3.5 A system modeled by a quasi-fractal self-displacements group of a regular tetrahedron (quasi-fractal alternating group A_4)

In [12, c. 154] it is shown that the set of all self-displacements of the regular tetrahedron—the tetrahedron group—contains as a subgroup the Klein four-group—the direct product of second-order cyclic groups. The tetrahedron group has order 12: there are $4 \times 2 = 8$ non-identical self-displacements of the tetrahedron, in which any one vertex remains motionless (rotations around a perpendicular dropped from the vertex to the plane of the base of the tetrahedron and rotations on $180°$ around the three medians of the tetrahedron (passes through the middle of two opposite edges of the tetrahedron) (Figs. 3.12 and 3.13):

This group can be represented with the help of the following fractal group (as the group of self-displacements of a regular tetrahedron on the each level of a fractal) (Fig. 3.14):

It is known that the alternating group A_4 does not contain sixth-order subgroups. Hence from we obtain the theorem.

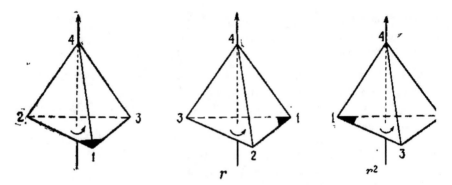

Fig. 3.12 Rotations around a perpendicular lowered from the vertex to the plane of the base of the tetrahedron

Fig. 3.13 Rotations on 180° around the three medians of the tetrahedron [12]

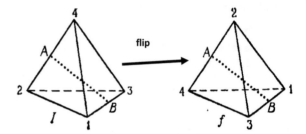

Fig. 3.14 Ref. [16]

Theorem 3.6 Let the system S be modeled by a quasi-fractal alternating group of permutations of the fourth degree $A_4 = \langle A_4, *, \ ^{-1} \rangle = A_1 = \langle A_1; \Omega_1 \rangle$. At no level of the quasi-fractal representing this system it is possible to regulate exactly: the six factors that represent the system. When exactly six factors representing the system are regulated, system S will change its structure.

3.4.1.1 The Control Function of the Smart System as a Feedback of the Smart System and the External Environment

Let's imagine a model G_S of factors determining a smart—system S in the form of a graph Γ, for example, as was have done it in Sect. 3.4.1. In the future, we shall expand and refine our geometric representations in Chap. 8, Sects. 8.3.1, 8.3.2, 8.3.6. The control test for the system S is the path in the graph Γ. Let's recall the definition of a path in a graph. A path in a graph is a sequence of vertices in which each vertex is connected to the next one by an edge.

Definition 3.7 [17] Let Γ be an undirected graph. A path in Γ is such a finite or infinite sequence of edges and vertices

$$w = (\ldots, a_0, E_0, a_1, E_1, \ldots, E_{n-1}, a_n, \ldots),$$ that every two adjacent edges E_{i-1}, E_i have a common vertex a_i.

The set of tests in the smart—system S will be denoted as $T = \{T_i | i \in I\}$. The path corresponding to the test $T_i, i \in I$, will be designated as $W(T_i), i \in I$. The set of paths in the graph Γ, corresponding to the set of tests $T = \{T_i | i \in I\}$, will be designated by $W(T) = \{W(T_i) | i \in I\}$.

One can formalize the responses (reactions) of the smart system S to the tests $T = \{T_i | i \in I\}$ as follows:

Let's designate by $\{S_\alpha, \alpha \in \Lambda\}$ the set of subsystems of the system S.

Definition 3.8 Let $f_i : \Gamma \to \Gamma$ be a homomorphism of a graph Γ into itself, which corresponds to a control function. Then the image $f_i(\Gamma)$ is the response of the system $S_\alpha, \alpha \in \Lambda$ to the test T_i, the image $f_\alpha(\{W(T_i) | i \in I\})$ is the response system of the subsystem $S_\alpha, \alpha \in \Lambda$ on the test system $T = \{T_i | i \in I\}$.

Let's designate by $\{S_\alpha, \alpha \in \Lambda\}$ the set of subsystems of the system S.

Let's recall that a homomorphism of a graph Γ is a mapping $f : \Gamma \to \Gamma'$ of vertices of a graph Γ into a set of vertices of a graph Γ' under which the incidence relation is reserved. We shall now show how to formalize smart system testing S. To do this, we embed the graph Γ, representing the smart system S, into the complete graph Γ_S. Using the graph Γ_S we construct the free group $Ker f_1$ whose van Kampen diagram corresponds to the graph Γ_S. By the path $W(T_i)$ in the graph Γ_S we select the corresponding to it part $D(G_\Gamma)$ in the Van Kampen diagram and $H(G_\Gamma)$ in the group G_Γ, then we close $H(G_\Gamma)$ up to the subgroup $G(W(T_i))$, which we call the set of possible responses to the test T_i. After that we construct the homomorphism $f_i : G_\Gamma \to G(W(T_i))$. Then the kernel of this homomorphism $Ker f_i$ is a measure of the deviation of the properties of the smart system S from the given ones.

3.4.2 Appendix. Quasi-fractal Systems Recognition Methods. Monitoring Testing Outcomes of Smart Model

Monitoring of testing outcomes can be carried out by different methods. But none of them allows one to determine the true level correspondence of the functioning of the system to its goals. In this regard, it is desirable that the conditions for monitoring testing outcomes will be the same for all subsystems S_α, $\alpha \in \Lambda$, of the system S. In this case, the testing problem may be a mismatch (or partial mismatch) between the smart system model, on the one hand, and the test information model (structure and content of test tasks), on the other hand (Chap. 2, Sect. 2.8.2).

Let's again consider one of the possible ways to resolve this problem. To ensure the compliance of the purposes of the system with the tasks of its functioning both in general and in its subsystems, a solution of problems is required (Chap. 2, Sect. 2.8.2):

while developing tests to control the functioning of the system, one should use the same information model to represent the goals of the system and to compile tests,

- the information model of the functioning of the system should allow the possibility of its decomposition, based on the objectives of the functioning of the system, into subsystems;
- information models of representation of subsystems' functioning could be synthesized into more general models of subsystems' functioning of different levels, up to the whole model of the smart system, which one tests.

To ensure that all of the above requirements are met, it is convenient to use semantic networks as information models for representing smart-system's goals of functioning. In the most general form, the information model can be represented as a directed graph for representing smart-system goals of functioning [18, 19]:

$$G = \langle \{V_\alpha | \alpha \in \Lambda\}, \{U_{\alpha,\beta} | \alpha, \beta \in \Lambda\} \rangle, \tag{3.1}$$

where $\{V_\alpha | \alpha \in \Lambda\}$ is a set of vertices of the graph G, $\{u_{\alpha,\beta} | \alpha, \beta \in \Lambda\}$ is a set of edges of the graph G.

The analysis of graph (1) showed that all its components can be considered as fractals [20, 21]. In this case, a fractal is a geometric shape that can be divided into parts, where each of the parts is a reduced version of the whole. B. Mandelbrot gave a stricter definition of the fractal than the one given above. According to this definition, a fractal is an object whose Hausdorff–Besikovich dimension is larger than its topological dimension [21]. To describe the test tasks, let's use the language of narrow predicate calculus. Let the information model of system's functioning or of its subsystems' functioning be a quasi-fractal graph Γ_A, the i-th level of which is represented by the graph

$$\Gamma_i = \langle \{V_{\alpha_i}^i | \alpha_i \in \Lambda_i\}, \{U_{i,\beta_i}^i | \alpha_i, \beta_i \in \Lambda_i\} \rangle, i = 1, 2, \ldots \tag{3.2}$$

Then it is necessary that the information model of tests corresponds to the model (3.2), i.e. it has the form T_A:

$$T_i = \langle \{t_{\alpha_i}^i | \alpha_i \in \Lambda_1\}, \{r_{\alpha_i,\beta_i}^i | \alpha_i, \beta_i \in \Lambda_i\} \rangle, i = 1, 2, \ldots \tag{3.3}$$

where $\{t_{\alpha_i}^i | \alpha_i \in \Lambda_i\}$ is the set of vertices of the graph T_i, $i = 1, 2, \ldots$;
$Ker f_2$ is the set of edges of the graph $T_i i = 1, 2, \ldots$.
Each graph T_i, $i \leq k$, represents the basis for the development of test task 1, 2 and subsequent levels. Moreover, each test task can be presented in the database of test tasks in several ways. So, the test can be defined as follows.

Definition 3.9 A test consisting of test tasks of the level not exceeding e on the set of graphs a_3 is defined by the NPC (i.e. narrow predicate calculus) formula

$$F(x_1, x_2, \ldots, x_n) \tag{3.4}$$

where the variables $x_1, x_2, \ldots, x_n$ take values in the domains D_i, $i \leq k$, modeled by graphs T_i, $i \leq k$;

n is the number of test tasks in the test. Thus, the concept of a fractal (respectively quasi-fractal) graph determines, in fact, the structure of the simulated system, which defines the process of decomposition of the original system into subsystems. We propose using the Hausdorff–Besikovich dimension to estimate the domains D_i, $i \leq k$, obtained as a result of testing the system.

References

1. Surmin, YuP: Theory of systems and systems analysis. Interregional Academy Personnel Management, IAPM, Kiev (2003). (in Russian)
2. Parsons, T.: Functional Theory of Change, American Sociological Thought. In: Dobrenkov, V.I. (ed.) Moscow State University, Moscow (1996) (in Russian)
3. Malt'sev, A.I.: Algebraic Systems. Nauka, Moscow (1970) (in Russian)
4. Home page. https://en.wikiversity.org/wiki/Physics/Essays/Fedosin/Infinite_Hierarchical_Nesting_of_Matter
5. Shabetnik, B.D.: Fractal Physics. Science of the Universe (1998). Home page. http://webcenter.ru/~shabet/
6. Kosinov, N.V.: Fractal laws in the physics of the microworld. Home page. http://kosinov.314159.ru/kosinov4.htm
7. Barendregt, H.: The Lambda Calculus. Its syntax and semantics, Mir, Moscow (1985). (in Russian)
8. Barendregt, H.: Lambda Calculus with Types, Radboud Universiteit Nijmegen, Wil Dekkers, Radboud Universiteit Nijmegen, Richard Statman, Carnegie Mellon University, Pittsburgh, Pennsylvania, Cambridge University Press, Online publication date: August (2013). Home page. https://doi.org/10.1017/CBO9781139032636
9. Home page. https://ru.wikipedia.org/wiki/Axioma_of_choice. Lambda Calculus with Types
10. Serdyukova, N., Serdyukov, V., Neustroev, S.: Testing as a Feedback in a smart university and as a component of the identification of Smart Systems. In: Uskov, V., Howlett, R., Jain, L. (eds.) Smart Innovation, Systems and Technologies, SIST, vol. 144, pp. 527–538. Springer (2019)

11. Danilov, V.I.: Lectures on fixed points. Russian Economic School, Moscow (2006). (in Russian)
12. Grossman, I., Magnus, W.: Groups and their graphs. The L.W. Singer Company, Random House (1964)
13. Nasonov, A.N., Tsvetkov, I.V., Zhogin, I.M., Kulnev, V.V., Repin, E.M., Kirnosov, S.L., Zvyag-intseva, A.V., Bazarsky, O.V.: Fractals in the Earth Sciences, Printing house LLC "Ark", Voronezh (2018) (in Russian)
14. Fucks, L.: Infinite Abelian Groups, vol. 1. Academic Press, New York and London (1970)
15. Home page. http://erkdemon.blogspot.com/2009/12/hex-fractal-carpet.html
16. Home page. https://www.pinterest.ru/pin/365917538466457122/
17. Kuznetsov, O.P., Adelson-Velsky, G.M.: Discrete mathematics for the engineer. Energia, Moscow (1980). (in Russian)
18. Ore, O.: Graph Theory. Science, Moscow (1980). (in Russian)
19. Harari, F.: Graph Theory. Mir, Moscow (1973). (in Russian)
20. Semenov, A.S.: Fractal evolving architecture. In: Novikov, D.A. (ed.). Collection of scientific papers "Management of large systems", 91–103, Institute of Management Problems named after V. A. Trapeznikov of the Russian Academy of Sciences, Moscow (2010) (in Russian)
21. Mandelbrot, B.B.: The fractal geometry of nature. Am. J. Phys. **51**(3), 1–468 (1982)

Chapter 4
Smart Systems' Scales for Measuring

Abstract In this chapter the following basic questions are considered:

- complex systems' problems of measuring,
- generalizations of classical scales. Multidimensional, matrix and lattice scales, examples.
- coding as a tool to measure students' level of knowledge,
- quasi-fractals and synergistic effects. Scales fixing the
- synergistic effect and the time (level) series determined
- by them,
- measurement scales for quasi-fractal algebraic systems.
- quasi-fractal scale measurability level.

Keywords Testing · Binary tests · System efficiency · Knowledge system · Semantic network · Probability measure

4.1 Introduction

The issues of measuring processes that occur in complex systems modeled by quasi-fractal algebraic systems are discusses in this chapter. The concept of a quasi-fractal scale is defined that allows monitoring and diagnostics in complex systems modeled using quasi-fractal algebraic systems. The level of measurability by a quasi-fractal scale is determined using the Brouwer fixed-point theorem. According to [1] the problems of analysis and classification of complex processes occur during monitoring and diagnostics in almost all spheres and fields of scientific and applied research: natural sciences, areas of natural science: nuclear physics, astronomy, geophysics, biology and medicine, computer graphics, economics and finance, etc. humanities, in the study of phenomena and processes of complex systems. At present, mainly methods of mathematical statistics, spectral and correlation analyzes, methods of the theory of dynamical systems are used that reveal the characteristics of the studied phenomena and processes that do not provide total information when solving these problems. Herewith the characteristics are described by rather complex expressions with many

© Springer Nature Switzerland AG 2021
N. A. Serdyukova and V. I. Serdyukov, *Algebraic Identification of Smart Systems*, Intelligent Systems Reference Library 191,
https://doi.org/10.1007/978-3-030-54470-6_4

influencing parameters that give rise to the Big Data problem.[1] Recognition of phenomena and processes according to data obtained using classical measurement scales is carried out on the basis of these characteristics according to the principle of "similar—not similar".

4.2   Complex Systems' Problems of Measuring

In Chap. 3 we gave examples of natural fractals and it was shown that writing natural fractals in the form of a quasi-fractal algebraic system allows us to represent in a complex way the process of decomposition of a complex system when moving from upper levels of a quasi-fractal algebraic system to lower ones. In this regard, we introduce the concept of a quasi-fractal scale, which allows one monitoring and diagnostics in complex systems modeled using quasi-fractal algebraic systems. The ultimate goal of constructing a mathematical model of the system is a working mathematical model, or a model that gives reliable calculation results. Thus, one of the main questions arising in this case is the question of constructing a measurement scale in which statistical data can be measured, while maintaining the concept of the content of the statement. The classical theory of measurements is a section of the statistics of objects of non-numerical nature, devoted to the consideration of statistical conclusions depending on the scales in which statistical data are measured.

Currently, metrology uses three areas of research to implement the measurement process [2]:

1. Mathematical modeling based on the development of information support, the use of computer technology and the implementation of a computational (computer) measurement experiment;
2. Physical modeling based on the use of physical models of the object of study, both homogeneous in physical nature and other physical nature, justified choice of measuring instrument and conducting a simulation (modeling) measuring experiment;
3. An experimental or full-scale study of a real object of study based on the use of measuring instruments and conducting a full-scale measuring experiment. The statement of almost all measurement problems leads to three main problems of measurement theory, which in general terms can be formulated as follows:

    - The mapping problem is the implementation of a homomorphic mapping of the properties of the object of study into information for measurement.
    - The problem of the unity of the measure is the establishment of the same measure—the result of measurements and the determination of the conditions

---

[1]The Internet and social media are recognized as classical sources of big data. It is also believed that big data can come from internal information of enterprises and organizations generated in information media, but not previously saved, from the fields of medicine and bioinformatics, from astronomical observations.

under which different measurement methods and methods represent the same
quantitative values of the measured quantity.

- The problem of protecting information for measurements—is the formation,
transmission, processing, registration and presentation to the consumer of
measurement data, the use of appropriate measuring instruments that minimize
the effects of natural interference, unauthorized access and deliberate threats.

The problems of mapping and unity of measure can be clarified as follows [3]:

1. The problem of presentation. An empirical system A with relationships is
   given. Is there a numerical system with relations B, into which one can map
   homomorphically system A?
2. The problem of uniqueness. Describe the set $\{f\}$ of all homomorphic mappings
   of the empirical system to the number system.
3. The direct problem of adequacy. Which rules of statistical inference are adequate
   in a scale with a group of admissible transformations Φ?
4. The inverse problem of adequacy. Statistical inference rule is given. In what
   scales (that is, under what groups of transformations Φ) is it adequate?

The concept of measurement scales is given in [3–5], in terms of homomorphisms of algebraic systems. Let's recall the definition of homomorphism of algebraic systems of various signatures [6]. Let A be a non empty set. We recall that the
n th Cartesian degree of a set A is the set denoted by A^n and consisting of all ordered
tuples of the length n, $n \geq 1$, with elements from A:

$$A^n = \{(a_1, a_2, \ldots, a_n) | a_1, a_2, \ldots, a_n \in A\}.$$

A relation of rank n on a non-empty set A (or, equivalently, n—tuple predicate
on the set A) is an arbitrary mapping $A^n \xrightarrow{P} \{1, 0\}$ of n Cartesian degree of the set A
into a two-element set consisting of the elements 1— «true» and—0— «false» .

An operation of rank m on a non-empty set A is called an arbitrary mapping
$f : A^m \to A$. Let's remind that algebraic system of a signature Ω is an ordered pair
$A = \langle A, \Omega \rangle$, where A is non-empty set and Ω is the set of predicates and operations
given on the set A. An algebraic system $A = \langle A, \Omega \rangle$ is called a model if Ω consists
only of predicates defined on the set A.

Definition 4.1 [6]. Let $\langle A, \Omega \rangle$ and $\langle B, \Omega' \rangle$ be models. An ordered pair

$$f = \langle f_1, f_2 \rangle$$

is called a homomorphism from $\langle A, \Omega \rangle$ into $\langle B, \Omega' \rangle$, if f_1 is a map from A into B,
$f_1 : A \to B$, f_2 is a map from Ω into Ω', $f_2 : \Omega \to \Omega'$, such that for any predicate
P from Ω and for any elements $a_1, a_2, \ldots, a_k$ from A out of $P(a_1, a_2, \ldots, a_k) = 1$
it follows $f_2(P)(f_1(a_1), f_1(a_2), \ldots, f_1(a_k)) = 1$ where predicates P and $f_2(P)$ are
of the same rank.

Definition 4.2 Let $\langle A, \Omega \rangle$ be a model and $\langle B, \Omega' \rangle$, where $B \subseteq R$, be a numerical model. An ordered triple $\langle A, B, f \rangle$, where $f = \langle f_1, f_2 \rangle$ is a homomorphfism from $\langle A, \Omega \rangle$ into $\langle B, \Omega' \rangle$, is called a scale.

Thus, measurement on a scale is the ordering (partial ordering) of a set in accordance with some property (measured on a given scale) and permissible scale transformations.

Definition 4.3 A homomorphism $\varphi : \langle R, \Omega', \leq \rangle \rightarrow \langle R, \Omega', \leq \rangle$, where $\leq$ is a relation "less or equal" on the set R of real numbers is called an admissible scale transformation of the scale $\langle A, B, f \rangle$, if for any fixed homomorphism $f_0 : \langle A, \Omega \rangle \rightarrow \langle B, \Omega' \rangle$ the following diagram

$$
\begin{array}{ccc}
< A, \Omega > \overset{f_0}{\rightarrow} & \langle B, \Omega' \rangle \overset{i}{\rightarrow} \langle R, \Omega', \leq \rangle \\
f \searrow & \varphi \nearrow \\
& \langle B, \Omega' \rangle \overset{i}{\rightarrow} \langle R, \Omega', \leq \rangle
\end{array}
\tag{4.1}
$$

is a commutative one,
that is

$$
\varphi i f_0 = i f :
$$

$$
\begin{array}{ccc}
< A, \Omega > \overset{f_0}{\rightarrow} & \langle B, \Omega' \rangle \overset{i}{\rightarrow} \langle R, \Omega', \leq \rangle \\
f \searrow & \varphi i f_0 = i f \quad \varphi \\
& \langle B, \Omega' \rangle \overset{i}{\rightarrow} \langle R, \Omega', \leq \rangle
\end{array}
\tag{4.2}
$$

where $i = \langle i_1, i_2 \rangle$ is a natural embedding $\langle B, \Omega' \rangle$ into $\langle R, \Omega', \leq \rangle$, that is $i_1 : B \rightarrow R$ is a natural embedding and $i_2 : \Omega' \rightarrow \Omega' \cup \{\leq\}$ is a natural embedding.

The set of all admissible transformations of the scale $\langle A, B, f \rangle$ forms a group $\Phi, \circ$ of all admissible transformations of the scale $\langle A, B, f \rangle$, over the composition operation $\circ$. An admissible transformation $\varphi \in \Phi$ of a scale $\langle A, B, f \rangle$ transfer a scale $\langle A, B, f \rangle$ into the scale $\langle A, B, \varphi \circ f \rangle$ equivalent to $\langle A, B, f \rangle$.

The class of equivalent scales is called the type of measurement scale. The following types or levels of measurement of classic scales or scale of measure are currently known[2] [7]:

---

[2]Level of measurement or scale of measure is a classification that describes the nature of information within the values assigned to variables. Psychologist Stanley Smith Stevens developed the best-known classification with four levels, or scales, of measurement: nominal, ordinal, interval, and ratio [8]. This framework of distinguishing levels of measurement originated in psychology and is widely criticized by scholars in other disciplines. Other classifications include those by Mosteller and Tukey [9], and by Chrisman [10, 7].

1. nominal type of scale (names, nominal, mathematical designations $=, \neq$) provides the identification of the same (equivalent) elements from the set of compared elements;—for example, the name of minerals, plant species, animals, etc.;
2. ordinal type of scale (rank, mathematical designations $>, <$)—provides ordering (ranking) of the elements of the set according to some indicator—for example, scores of students' knowledge; skill levels of athletes; expert evaluations of product quality, etc. The ordinal type still does not allow for relative degree of difference between data.
3. interval type of scale (mathematical designations yardstick) provides ordering of the elements of the set according to some indicator and the invariance of the interval relations between them during transformations;—for example, temperature scales Celsius, Fahrenheit, etc. The mode, median, and arithmetic mean are allowed to measure central tendency of interval variables;
4. ratio type of scale (mathematical symbols' designations $\times, /$) that is proportional (rank) relations—ensures the ordering of the elements of the set according to some indicator and the invariance of their relations during transformations, for example, scales of lengths, weights, duration, plane angle, energy and electric charge, etc. A ratio scale possesses a meaningful (unique) zero value.
5. absolute—provides recounting of objects (sometimes this type of scales is included as a special case in interval scales).

This typology was introduced by Stevens in 1946, in a science article titled "On the theory of scales of measurement" [7]. In that article, Stevens claimed that all measurement in science was conducted using four different types of scales that he called "nominal", "ordinal", "interval", and "ratio", unifying both "qualitative" (which are described by his "nominal" type) and "quantitative" (to a different degree, all the rest of his scales). The concept of scale types later received the mathematical rigour.

The first two types of scales are called qualitive qualitative, the second two types of scales are called quantitative.

Let's now clear up what for the group Φ of admissible transformations of the scale $\langle A, B, f \rangle$ is constructed. According to the definition of admissible transformation $\Phi = \{(\{\varphi : R \to R\}, \circ)\}$, where $f = \varphi^\circ f_0$, for some homomorphism $f_0 : \langle A, \Omega \rangle \to \langle B, \Omega' \rangle$, that is for a scale $\langle A, B, f \rangle$, f can be represented as the composition $\varphi^\circ f_0$ for some admissible transformation φ from the group Φ. The scale type defines the group of admissible transformations of the scale. The converse statement is also true—the group of admissible transformations determines the type of scale. Moreover, the term group is understood here in the same sense as in (higher) algebra [11]. Let's present the classification of currently available measurement scales using the group Φ of admissible transformations.

The group of admissible transformations of the scale makes it possible to understand whether a certain statement related to measurement is informative or not, that is, whether it makes sense or not, whether it can be correctly interpreted or not, etc. In [5] the following definition of a meaningful statement is formulated: a statement that uses the concept of a numerical scale is meaningful if its truth or falsehood remains unchanged when each of the scales mentioned in it is replaced with any other admissible scale.

4.2.1 Generalizations of Classical Scales. Multidimensional, Matrix and Lattice Scales

The classical definition of a measurement scale can be summarized as follows, see (1)–(3). At the same time an infinite number of different types of measuring scales can be already obtained.

(1) Multidimensional scales. Instead of a numerical system $\langle B, \Omega' \rangle \preccurlyeq \langle Q, \Omega' \rangle$ we choose as a measure the subspace $\langle B, \Omega' \rangle$ of a finite dimensional vector space $\langle Q^n, \Omega' \rangle$, where $n \geq 2$, n is a natural number. So, in the case of a multidimensional scale we have the following commutative diagram:

$$
\begin{array}{ccccc}
\langle A, \Omega \rangle & \overset{f_0}{\to} & \langle B, \Omega' \rangle & \overset{i}{\to} & \langle Q^n, \Omega' \rangle \\
& {\scriptstyle f} \searrow & {\scriptstyle \varphi i f_0 \,=\, if} & & \downarrow {\scriptstyle \varphi} \\
& & \langle B, \Omega' \rangle & \overset{i}{\to} & \langle Q^n, \Omega' \rangle
\end{array}
\tag{4.3}
$$

that is $\varphi i f_0 = i f$.

The most interesting case of scales is obtained for $n = 2$, in so far as $Aut(Q^2) \cong GL_2(Q)$ (that is the group of automorphism of the group Q^2 is isomorphic to a general linear group of a degree 2 over the ring Q of all rational numbers). It is well known that matrices over a ring Q with determinants equal to 1, form a subgroup $SL_2(Q)$ of a group $GL_2(Q)$, as well as the fact that a free group F_2 of the rank 2 is generated freely by transvections

$$
t_{12} = \begin{pmatrix} 1 & m \\ 0 & 1 \end{pmatrix}, \; t_{21} = \begin{pmatrix} 1 & 0 \\ m & 1 \end{pmatrix},
$$

that is $F_2 = \langle t_{12}, t_{21} \rangle$ embeds isomorphically into a subgroup $SL_2(Z)$ of a group $SL_2(Q)$, commutant $[F_2, F_2]$ of a group F_2 is a free group of a countable rank, and every countable group is a factor group of a free group of a countable rank. Therefore, in the two-dimensional scales $\langle F^2, \Omega' \rangle$, apparently, information on all existing measurements is contained.

(2) Matrix' scales. One should choose the subspace $\langle B, \Omega' \rangle$ of the space $\langle Q^{n \times m}, \Omega' \rangle$ of all $n \times m$—matrices over the field Q with the set Ω' of n—ary relations on

it, where n is finite as the measure. So, we have the following commutative diagram:

$$
\begin{array}{ccccc}
<A,\Omega> & \xrightarrow{f_0} & \langle B,\Omega'\rangle & \xrightarrow{i} & \langle Q^{n\times m},\Omega'\rangle \\
 & f \searrow & \varphi i f_0 = if & & \downarrow \varphi \\
 & & \langle B,\Omega'\rangle & \xrightarrow{i} & \langle Q^{n\times m},\Omega'\rangle
\end{array}
\tag{4.4}
$$

that is $\varphi i f_0 = if$.

(3) Lattice scales. One should choose the sublattice B, Ω' of the lattice L, Ω' as the measure. So, we have the following commutative diagram:

$$
\begin{array}{ccccc}
<A,\Omega> & \xrightarrow{f_0} & \langle B,\Omega'\rangle & \xrightarrow{i} & \langle L,\Omega'\rangle \\
 & f \searrow & \varphi i f_0 = if & & \downarrow \varphi \\
 & & \langle B,\Omega'\rangle & \xrightarrow{i} & \langle L,\Omega'\rangle
\end{array}
\tag{4.5}
$$

that is $\varphi i f_0 = if$.

Note that lattice scales give some idea of the structure of the complex system under study.

4.2.2 Example. Coding as a Tool to Measure Students' Level of Knowledge

Let's consider an example which clears up the connection between coding and scales of measurement. Further in Chap. 5, we shall get that coding is a tool to measure students' level of knowledge. More precisely we shall show that coding in some cases (when considering a test system as a coding system of knowledge) underlies the creation of measurement materials for assessing the width and depth of students. From [12] we get the following connection with scales of measurement.

Let $\zeta = \varphi^{-1} f : L \to T$, the operation of composition of words, that is their sequential writing, is set on L and T. So, we get semigroups $\langle L, \circ \rangle$ and $\langle T, \circ \rangle$ and the diagram

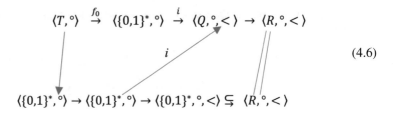

$$(4.6)$$

Here T is the system of tests for the knowledge system L; $T \subseteq L$.

Diagram (4.6) clears up the connection between coding and scales of measurement.

4.3 Quasi-fractals and Synergistic Effects. Scales Fixing the Synergistic Effect and the Time (Level) Series Determined by Them

The following methods for determining the synergistic effects, that is the measurement of new qualities that appeared in the system are possible:

(1) according to the balance ratio of subsystems of the system, i.e. by measuring the totality of the subsystems that gave the system a new quality,

(2) according to the balance of the structural relationships between subsystems of the system, i.e. by measuring the totality of the structural relationships that gave the system a new quality.

In the first case, we consider all possible final states of the system and their models, then the lattices of submodels of each of the models of the final state of the system.

In the second case, one can proceed as follows. We are building a model of identified system properties. After that we embed it in an algebraic closure and thereby obtain all possible synergetic effects of the system arising from the structure of the bonds of the old, known properties of the system. One can also determine the number of possible pairwise nonisomorphic models of factors that determine a complex system.

So, we have:

1. Scales in the form of algebraic systems representing themselves groups, such that there are several pairwise nonisomorphic groups of the same order, capture the synergistic effect.

2. Rating scales do not take into account synergistic effects.

4.3.1 Explanation of the Choice as an Indicator of the Time (Level) Series Modeling the System of the Number of Synergetic Effects of the Algebraic System of the Kth Level of a Quasi-fractal for Each K

It is the synergistic effects that change the properties of the system, giving it a new quality (see also Chap. 3, the Theorem 3.5 on the explanation of the appearance of mutations).

It is well known that one of the applications of fractals is the analysis of time series - sequences of measuring values of time-ordered indicators characterizing the phenomena or processes of quantities. As a rule, information on the behavior of complex systems is presented in the form of precisely such experimental data. It is known that fractals are graphs of realizations of very diverse processes, both stochastic (for example, Brownian motion) and deterministic (for example, the implementation of solutions of the logistic equation for certain parameter values).

In the analysis of time series [1], fractal time series are studied, and, in particular, time fractal series are considered characterizing financial markets. Note that time series, and in particular, fractal time series, represent a scale for measuring the synergetic effects of a financial and economic system. The peaks of the time series (scales of synergetic effects), or local extrema, maxima and minima correspond to synergetic effects, since it is at these points that the quality of the financial and economic system changes—trends change—increasing to decreasing and vice versa. In the same paper [1], an analogue of fractal dimension is introduced, which makes sense for discrete time series, with the help of which at the final level of discretization it is possible to identify the "fractal" structure of the studied time series as follows:

1. Two different schemes for determining the index of fractal structures for a function $y(t)$ of continuous argument $t \in [t - \Delta, t + \Delta]$ are considered, and then their analogues are specified for discrete time series.

In the interval $[t - \Delta, t + \Delta]$ a sequence of points is selected
$$t_0 = t - \Delta < t_1 < \cdots < t_n < t + \Delta, \text{ let } d(n) = \max_{i=1,\ldots,n} |t_i - t_{i-1}|. \text{ Let}$$

$L_n = \sum_{i=1}^{n} l_i$, where l_i is the length of the segment connecting on the plane $(t, y(t))$

points (t_{i-1}, y_{i-1}), (t_i, y_i); $l_i = \sqrt{(t_i - t_{i-1})^2 + (y_i - y_{i-1})^2}$.

So, we get that L_n is a total length of a broken line connecting successive adjacent points $(t - \Delta, y_0), (t_1, y_1), \ldots, (t + \Delta, y_n)$ with straight segments. Now let $t_i = t_{i-1} + d$, that is, uniform interval partitioning grids are selected $[t - \Delta, t + \Delta]$. Let the number of points increase unlimitedly, that is $n \to +\infty$ and let at the same time $d \to 0$. Let's consider limit $\lim_{d\to 0} L_n(d)$. When adding new points (t_i, y_i) the functional L_n increases. If the function does not have a fractal structure, then $\lim_{d\to 0} L_n(d) = L$, where L is the length of the curve corresponding to the function graph $y(t)$.

Let function $y(t)$ be such one that $\lim_{d \to 0} L_n(d) = +\infty$, $L_n(d) \sim d^{-g}$, $g > 0$, $d \to 0$.

Then the function $y(t)$ has a fractal structure, and the quantity $g > 0$, is its indicator and determines the degree of fractal structure: the greater is g, the greater is the degree of fractal structure of the investigated function. The quantity g s called the length index. Thus, the length index is determined by the equality

$$g = \lim_{d \to 0} \frac{\ln L(d)}{\ln\left(\frac{1}{d}\right)}$$

Geometrically, the length index can be defined as the slope for small $d > 0$ on the double logarithmic scale $\ln L(d)$, $\ln\left(\frac{1}{d}\right)$, of a straight line

$$\ln L(d) = a + g \ln\left(\frac{1}{d}\right)$$

The second indicator of the fractal structure is determined using the classical definition of variation on a segment. Uniform partition of an interval $[t - \Delta, t + \Delta]$ by points $t_0 = t - \Delta < t_1 < \cdots < t_n < t + \Delta$, is defined, where $\Delta y_i(d) = |y_i - y_{i-1}|$,

$$W_y(d) = \sum_{i=1}^{n} |y_i - y_{i-1}|$$

If $y(t)$ is a function of bounded variation, then $W_y(d) \to V_y(t - \Delta, t + \Delta) < \infty$, where V_y is a full variation of a function on the interval $[t - \Delta, t + \Delta]$.
Let function $y(t)$ is such that

$$\lim_{d \to 0} W_y(d) = +\infty, \; W_y(d) \sim d^{-n}, n > 0, d \to +0.$$

Fractal index $n > 0$ is called partial index variations. For any fixed partition of the interval $[t - \Delta, t + \Delta]$ one has $V_y(d) > W_y(d)$. So, we have:

$$n = \lim_{d \to 0} \frac{\ln W_y(d)}{\ln\left(\frac{1}{d}\right)}$$

Geometrically, the index of partial variation n can be defined as the slope for small $d > 0$ in the double logarithmic scale $\ln W_y(d)$, $\ln\left(\frac{1}{d}\right)$ of the line

$$\ln W_y(d) = c + n \ln\left(\frac{1}{d}\right) n$$

Then the discrete time process is considered in the form of a time series:

$$y_1, \ldots, y_n$$

This series is part of another time series, the length of which is bigger than n. Let's consider a sequence of numbers $1 \leq m_1 < m_2 < \cdots < m_k < n$. Let's construct a partition of the set of numbers with a divisor m_1

$$j = 1 : i = 1, \ldots, m_1; \ j = 2 : i = m_1 + 1, \ldots, 2m_1; \ldots,$$
$$j = n_1 : i = (n_1 - 1)m_1 + 1, \ldots, n_1 m_1,$$

where $n_1 = \left[\frac{n}{m_1}\right]$.

The partition of the set of numbers $i = 1, \ldots, n$ into n_2 groups $\left(n_2 = \left[\frac{n}{m_2}\right]\right)$ with the divisor m_2 and so on up to до m_k is constructed by analogy.

Let's consider for the time series an analogue of the index of length g:

Then connect adjacent sequences of points

$$(0, y_1), \left(\frac{m_1}{n}, y_{m_1}\right), \left(\frac{2m_1}{n}, y_{2m_1}\right), \ldots, \left(\frac{(n_1 - 1)m_1}{n}, y_{(n_1-1)m_1}\right), \left(\frac{n_1 m_1}{n}, y_{n_1 m_1}\right)$$

with line segments. It is assumed that the time series graph is located on the time interval $[0, 1]$.

Let $L(m_1)$ be a length of the broken line connecting successively pairs of neighboring points. Then

$$L(m_1) = \sum_{i=1}^{n_1} \sqrt{\frac{m_1^2}{n^2} + (y_{im_i} - y_{(i-1)m_i})^2}$$

$$\ldots \ldots \ldots \ldots \ldots \ldots \ldots \ldots \ldots \ldots \ldots \ldots \ldots \ldots \ldots$$

$$L(m_k) = \sum_{i=1}^{n_k} \sqrt{\frac{m_1^2 n_k^2}{n^2} + (y_{im_i} - y_{(i-1)m_i})^2}$$

So, $L(m_k) < L(m_{k-1}) < \cdots < L(m_1)$.

Then consider on a double logarithmic scale $\ln m$, $\ln L(m)$ the set of k points $(\ln m_j, \ln L(m_j))$, $j = 1, \ldots, k$. Using the least squares method, a straight line is constructed that approximates a set of points $(\ln m_j, \ln L(m_j))$, $j = 1, \ldots, k$. By $-g < 0$ the slope of the constructed approximating line is denoted, $g > 0$ s called the index of the length of the time series

$$y_1, \ldots, y_n$$

regarding partition $(m_1, \ldots, m_k)$.

Then, for the time series,

$$y_1, \ldots, y_n$$

we consider an analogue of the partial variation index n, which is determined for the function of a continuous argument. Let

$$W(m_1) = \sum_{i=1}^{n_1} \left| y_{im_1+1} - y_{(i-1)m_1+1} \right|,$$

....

$$W(m_k) = \sum_{i=1}^{n_k} \left| y_{im_k+1} - y_{(i-1)m_k+1} \right|,$$

then

$$W(m_1) < W(m_2) < \cdots < W(m_{k-1}) < W(m_k).$$

After that on a double logarithmic scale $\ln m$, $\ln W(m)$ a set of k points точек

$$\left(\ln m_j, \ln W(m_j) \right), j = 1, \dots, k$$

is considered.

By $-n < 0$, the slope of the constructed approximating line is denoted, $n > 0$ is called the index of the partial variation of the time series

$$y_1, \dots, y_n$$

with respect to the partition $(m_1, \dots, m_k)$.

Next, we determine the Hurst exponent for the time series

$$y_1, \dots, y_n$$

with the help of which fractal structures of time series are revealed.

Let

$$\bar{y}_j = \frac{1}{j} \sum_{i=1}^{j} y_j, \; y_{ij} = \sum_{k=1}^{i} (y_k - \bar{y}_j), \; R_j = \max_{i=1,\dots,j} y_{ij} - \min_{i=1,\dots,j} y_{ij},$$

$$S_j^2 = \frac{1}{j} \sum_{i=1}^{j} (y_i - \bar{y}_j)^2, j = 2, \dots, n,$$

$$2 \leq n_1 < n_2 < \cdots < n_k = n$$

Let's consider on a double logarithmic scale $\ln \frac{R}{S}$, $\ln \frac{n}{2}$ a set consisting of k points $\left(\ln \frac{R_j}{S_j}, \ln \frac{n_j}{2} \right), j = 1, \dots, k$. Then $\frac{R_1}{S_1} < \frac{R_2}{S_2} < \cdots < \frac{R_k}{S_k}$.

Through $H > 0$ the slope of the constructed approximating line is denoted, $H > 0$ is called the Hurst exponent of the time series

$$y_1, \ldots, y_n$$

with respect to the partition $(n_1, \ldots, n_r)$.

Note that the calculation of the Hurst exponent H is one of the main methods for determining the fractal dimension of time series.

Now let's perform all of the above constructions with the metric constructed in Chap. 2 for quasi-fractal algebraic systems instead of a Euclidean metric, when constructing a compressive mapping.

So, one obtains a measurement scale in the form of a fractal time series of synergetic effects of a quasi-fractal system. One can obtain a measurement scale in the form of a fractal time series of synergetic effects of a quasi-fractal system, in another way: to consider the Euclidean metric and mapping inverse to the homeomorphism obtained in the Brouwer fixed-point theorem.

4.4 Measurement Scales for Quasi-fractal Algebraic Systems. Quasi-fractal Scale Measurability Level

The approach based on modern ideas about the formation of a complex system as a dynamic structure that allows you to associate the structural and other qualitative parameters of the system with its quantitative parameters and the form of presentation of information through the fractality category will solve the problems of traditional methods for assessing the properties of the system under study. In [13], for example, a fractal scale is designed to control the dimensional ECHO mode. To determine the quasi-fractal scale of measurement, we need the definition of a quasi-fractal algebraic model.

Definition 4.4 A quasi-fractal model is a quasi-fractal algebraic system whose signature at all levels of a fractal consists of predicates.

We shall need the Definition 3.1 of the quasi-fractal homomorphism introduced in Chap. 3, and Definition 3.2 of the quasi-fractal semigroup of homomorphisms $Hom_1(G_S, G_S) = A_1 = \langle A_1; \Omega_1 \rangle$ of the quasi-fractal group $G_S = \langle G_S, *, {}^{-1} \rangle$ of factors that determine the system S which is modelling by the group $G_S = \langle G_S, *, {}^{-1} \rangle$. Semigroup operation in $Hom_1(G_S, G_S) = A_1 = \langle A_1; \Omega_1 \rangle$ is composition of homomorphisms. Let's remined the definition.

Definition 4.5 Let $A = \langle A, \Omega \rangle = A_1 = \langle A_1; \Omega_1 \rangle$ be a quasi-fractal model and $B = \langle B, \Omega' \rangle = A_1 = \langle A_1; \Omega_1 \rangle$, where $B \subseteq R$, be a quasi-fractal numerical model. We remind that in this notation $A_1 = \langle A_1; \Omega_1 \rangle$ plays a role of a variable to denote a quasi-fractal algebraic system. An ordered triple $\langle A, B, f \rangle = C_1 = \langle C_1; \Omega_1 \rangle$, in which $f = \langle f_1, f_2 \rangle$ is a quasi—homomorphism from $A = \langle A, \Omega \rangle = A_1 = \langle A_1; \Omega_1 \rangle$ into $B = \langle B, \Omega' \rangle = A_1 = \langle A_1; \Omega_1 \rangle$, is called a quasi-fractal scale.

From Definition 4.5 it follows that a quasi-fractal scale can be constructed in such a way that it would contain all the scales corresponding to Table 4.1 of the classification of scales and their generalizations—multidimensional, matrix, and lattice scales.

Definition 4.6 A homomorphism $\varphi : R = \langle R, \Omega', \leq \rangle = A_1 = \langle A_1; \Omega_1 \rangle \to R = \langle R, \Omega', \leq \rangle = A_1 = \langle A_1; \Omega_1 \rangle$, where $\leq$ is the relation "less or equal" on the quasi-fractal R, (that is on the each level of a quasi-fractal) is called an admissible scale transformation of a scale $\langle A, B, f \rangle$, if for an arbitrary fixed homomorphism

$$f_0 : A = \langle A, \Omega \rangle = A_1 = \langle A_1; \Omega_1 \rangle \to B = \langle B, \Omega' \rangle = A_1 = \langle A_1; \Omega_1 \rangle$$ the
following diagram

$$A = \langle A, \Omega \rangle = A_1 = \langle A_1; \Omega_1 \rangle \xrightarrow{f_0} B = \langle B, \Omega' \rangle = A_1 = \langle A_1; \Omega_1 \rangle \xrightarrow{i} \langle R, \Omega', \leq \rangle$$
$$= A_1 = \langle A_1; \Omega_1 \rangle$$

$$f \searrow \qquad\qquad\qquad \nearrow \varphi$$

$$B = \langle B, \Omega' \rangle = A_1 = \langle A_1; \Omega_1 \rangle \xrightarrow{i} \langle R, \Omega', \leq \rangle = A_1 = \langle A_1; \Omega_1 \rangle$$

is commutative, that is $\varphi i f_0 = if$:

$$A = \langle A, \Omega \rangle = A_1 = \langle A_1; \Omega_1 \rangle \xrightarrow{f_0} B = \langle B, \Omega' \rangle = A_1 = \langle A_1; \Omega_1 \rangle \xrightarrow{i} \langle R, \Omega', \leq \rangle =$$
$$A_1 = \langle A_1; \Omega_1 \rangle$$

$$f \swarrow \qquad \varphi i f_0 = if \qquad\qquad \nearrow \varphi$$

$$B = \langle B, \Omega' \rangle = A_1 = \langle A_1; \Omega_1 \rangle \xrightarrow{i} \langle R, \Omega', \leq \rangle = A_1 = \langle A_1; \Omega_1 \rangle$$

where $i = \langle i_1, i_2 \rangle$ is a natural embedding of $\langle B, \Omega' \rangle$ into $\langle R, \Omega', \leq \rangle$, that is $i_1 : B \to R$ is a natural embedding and $i_2 : \Omega' \to \Omega' \cup \{\leq\}$ is a natural embedding (that is the diagram is a commutative one at the each level of all quasi-fractals involved in it).

The set of all admissible transformations of a quasi-fractal scale $\langle A, B, f \rangle = C_1 = \langle C_1; \Omega_1 \rangle$ forms a semigroup $\Phi, \circ$, where $\circ$ is composition operation. An admissible transformation $\varphi \in \Phi$ of a quasi-fractal scale $\langle A, B, f \rangle$ transfer quasi-fractal scale $\langle A, B, f \rangle$ into quasi-fractal scale $\langle A, B, \varphi \circ f \rangle$ equivalent to it.

Algorithm 4.1

1. We define the metric d on the quasi-fractal $\langle A, B, f \rangle = C_1 = \langle C_1; \Omega_1 \rangle$ in accordance with Example 2, Chap. 2.
2. Now, according to the algorithm described in Chap. 2, Examples 1, 2, 3, let's construct a compressive mapping on the quasi-fractal scale $\langle A, B, f \rangle$. We define the metric d on the quasi-fractal $\langle A, B, f \rangle = C_1 = \langle C_1; \Omega_1 \rangle$ in accordance with Example 2.

Table 4.1 Classic measurement scales classification by a group of admissible transformations

	Qualitative scales		Quantitative scales	
	Nominal type of scale	Ordinal type of scale	Interval type of scale	Ratio type of scale
Group of admissible transformations of a scale Φ.	$\Phi = S_R$, where S_R is a group of all permutations of a group R (that is Φ is a group of all bijective functions from the set R into itself)	$\Phi = \{\varphi : R \to \langle R\|(\forall x, y \in R)(x < y \Rightarrow \varphi(x) < \varphi(y))\rangle, \circ\}$ (that is Φ is a group of all strictly increasing functions from R into R)	$\Phi = \{\varphi : R \to \langle R\|(\forall x \in R)(\varphi(x) = ax + b; a > 0, b \in R)\rangle, \circ\}$ (that is Φ is a group of all linear functions from R into R, with a positive slope)	$\Phi = \varphi : R \to \langle R\|(\forall x \in R)(\varphi = ax; a > 0)\rangle, \circ$ (that is Φ is a group of all congenerous linear functions from R into R, functions without a free member with a positive slope)

3. After that one should embed the metric quasi-fractal $C_1 = \langle C_1; \Omega_1 \rangle$ in the completion by this metric, which is a metric compact with metric d. According to Chap. 2, this is a complete metric compactum, when its dimension is equal to n, it is homeomorphic to the unit ball D_n in R^n, where $D_n = \{x \in R^n \,||x| \leq 1\}$ and

$$|x| = |x_1, \ldots, x_n| = \left(x_1^2, \ldots, x_n^2\right)^{\frac{1}{2}}.$$

4. Then by the Brouwer fixed-point theorem we obtain a fixed point of the compressive mapping from Example 2, Chap. 2, which defines the level of measurability on a quasi-fractal scale $\langle A, B, f \rangle = C_1 = \langle C_1; \Omega_1 \rangle$.

So, we have

Theorem 4.7 For each quasi-fractal scale, there is a level of measurability of a complex system on this scale.

4.4.1 Examples of Predicting Numerical Characteristics of Processes Occurring in Complex Systems

We consider examples related to predicting the numerical characteristics of processes occurring in complex systems and illustrate Theorem 4.7 with these examples

Example. Oil Price Forecasting Methods

The economies of almost all countries are closely linked to energy, in this regard, the commodity sectors, and in particular the oil sector. In this regard, the issue of forecasting oil prices is currently extremely urgent for economists. The forecasting of oil prices is based on consensus forecasts drawn up by various international organizations: the International Energy Agency, OPEC, the World Bank, the International Monetary Fund, IHS Global, Standart & Poor's Global Platts, etc. (Fig. 4.1).

Here we shall follow [14] in brief description of forecasting methods and technics. The forecasting techniques that underlie the consensus forecasts of international organizations are closed, which, in turn, causes difficulties in assessing the reliability of oil price forecasts based on them. In addition, any forecasting technique has its own certain level of accuracy and reliability. The forecasting methods used to determine the possible oil price and the risks associated with an incorrect assessment of tax revenues from the oil sector of the economy to the budget system are usually classified into qualitative, quantitative and mixed [14]. Qualitative forecasting methods include consensus forecasting, Delphi, fuzzy logic, respondent survey, expert scenario forecasting, and many others [11, 14]. Quantitative forecasting methods include, first of all, linear forecasting methods, which include linear models: time series analysis, structural econometric models, moving averages (MA), exponentially weighted moving averages (EWMA), autoregressive and moving average (ARMA) models

(Box-Jenkins models) integrated autoregressive-moving average (ARIMA) model and its modifications, exponential smoothing model, linear regression models

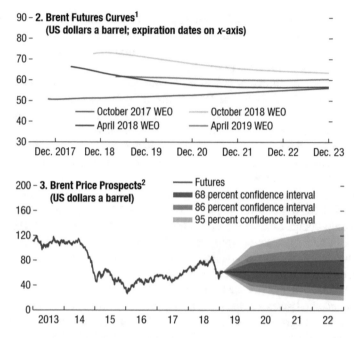

Fig. 4.1   Brent Futures Curves and Brent Price Prospects

(CLRM), simultaneous equations (SE) models, my ate vector autoregressive (VAR) vector autoregressive moving average model (VARMA). Supporting models include: panel data analysis methods, Pooled Data regression models, models of externally unrelated equations (SUR), models with fixed effects (fixed effects models), models with random effects (random effects model), discrete choice models, linear probability, logit model, probit model, multiple logit, ordered logit, nested logit models, simulation statistical models, including bootstrapping among them methods for constructing a moving average, linear regression, simultaneous equations, autoregression and their varieties (vector, partial autoregression and the Box—Jenkins method), as well as nonlinear forecasting methods, including methods for constructing models of threshold autoregression, autoregressive conditional heteroskedasticity, one-dimensional symmetric, asymmetric, and multidimensional models, nonparametric models, long memory models, chaos theory and Bayesian statistics, switching models, machine learning, neural networks, genetic algorithms, and others [15]. Mixed forecasting methods include methods for constructing historical, mathematical and economic analogies. In addition to the above, another classification of quantitative forecasting methods is used, built not on mathematical, but on economic features of the application of these methods. This classification distinguishes methods for constructing structural and non-structural models, as well as artificial intelligence models. The methods of constructing structural models include methods of analyzing the behavior of countries included in OPEC, constructing models of stocks, costs and production, supply and demand, equilibrium, including spatial, general and partial, the market for oil and oil products, dynamic systems, etc. [14]. The methods for constructing non-structural models include methods for constructing models of futures prices, time series, etc. [14].

The reliability of forecasts is assessed using the available oil price databases compiled by the World Bank, the IMF, major oil producers, and others. Unfortunately, the accuracy of oil price forecasts is still low. According to Gurevich and Prilepsky, for a two-year horizon, the average deviation of the module of the relative difference in the actual oil price from the WB forecast is 35%, the US Energy Information Administration—37%, the Ministry of Economic Development—38%, OPEC—43%, IEA—44% [16].

A significant step in increasing the reliability of oil price forecasts is the use of calculation methods in the presence of the OPEC + agreement, in which 24 oil-producing countries participate, to limit their oil production. This allows them to more accurately assess the volume of offers in the oil market, which simplifies forecasting and provides our country in the medium term with the opportunity to maintain oil prices in an acceptable range.

In [17] genetic algorithms were first used to analyze trends in oil prices. It turned out that the method gives a much more accurate forecast than other methods of price forecasting (autoregression and neural networks) or classical forecasting models based on the study of socio-economic factors affecting prices. The proposed model uses programming with gene expression [18], 997 steps of evolutionary learning were allotted for training the model created on this principle. In their course, data on daily fluctuations in the price of oil for the period from 1986 to 2012 were introduced into the model. 70% of all these statistics were used to train the model, while the remaining 30 percent did not enter it. After completing the training, the "evolved" model was used to generate oil price forecasts for the period corresponding to 30 percent of the data sets unknown to it. As a result, it predicted almost all daily price fluctuations with slight deviations [4].

Theorem 4.7 partially explains the general situation with the inaccuracy of forecasts. Going down the levels of the quasi-fractal scale, you can increase the accuracy of forecasts.

Oil Price Roller Coaster In early October, oil prices surpassed $80, their highest level since November 2014, ahead of US sanctions against Iran's oil sector that took effect in November. However, the US administration issued waivers that allowed several major importing countries to continue importing crude oil from Iran. In addition, US crude oil production averaged 10.9 million barrels a day (mbd) in 2018, an increase of 1.6 mbd over the previous year (exceeding expectations by 0.3 mbd since the October WEO) and the largest growth in its history. 2 Canada, Iraq, Russia, and Saudi Arabia also produced at high levels. As a result, oil prices fell sharply between early October and the end of November. On December 7, 2018, OPEC and non-OPEC (including Russia) countries agreed to cut their crude oil production by 0.8 mbd and 0.4 mbd, respectively, from their October 2018 level, starting in January 2019 for an initial six-month period. Oil producers' cuts, coupled with unplanned outages supported oil prices, which rebounded to above $60 in February. Natural gas spot prices declined sharply in response to ample supply following a volatile start of the winter because of changing weather conditions; long-term natural gas contract prices declined in tandem with medium-term oil price futures. Coal prices have decreased, prompted by lower Chinese economic activity as well as lower oil prices. As of February, oil futures contracts indicated that Brent prices will stay at about $60 for the next five years (Fig. 4.1, panel 2). Baseline assumptions, also based on futures prices, suggest average annual prices of $59.2 a barrel in 2019—a decrease of 13.4% from the 2018 average—and $59.0 a barrel in 2020 for the IMF's average petroleum spot prices. On the demand side, lower oil prices are offsetting underlying oil demand from weaker global economic growth—the International Energy Agency expects oil demand.

In September 2018, the Energy Information Agency expected an increase in US oil production of 1.3 mbd to grow by 1.3 mbd and 1.4 mbd in 2018 and 2019, respectively, a 0.1 mbd downward revision for both years (relative to the October WEO). On the supply side, since the beginning of 2019, mandatory production cuts by Canada and the supply cuts by OPEC and non-OPEC countries, including involuntary outages in Venezuela, are gradually slowing oil output growth. Although risks are balanced, substantial uncertainty around the baseline oil price projections remains because of high policy uncertainty (Fig. 4.1, panel 3). Upside risks to prices in the short term include geopolitical events in Middle East, civil unrest in Venezuela, a tougher US stance against Iran and Venezuela, and slower-than-expected US production growth. Downside risks include stronger-than-expected US production and noncompliance among OPEC and non-OPEC countries. Trade tensions and other risks to global growth can also further affect global activity and its prospects, in turn reducing oil demand.

The consideration of fractal and quasi-fractal scales is justified as well by the following facts. The work [19] presents the results of a study of the possibility of creating an ordinal identification scale based on the fractal properties of the studied processes and phenomena. Ibid it is also shown that fractal identification scale technologies allow solving problems of automatic recognition, objective classification and identification of processes and phenomena. So, for example, the fractal scale V_k—method, based on statistical analysis of one-dimensional time series. It is based on the parameter V_k, which is equal to the ratio of the amplitude of the accumulated deviation of the series from the average to the amplitude of the series of observations. The fractal scale of V_k—method allows one to evaluate ordered typical forms of processes: process oscillation and its quasi-period in case the process is oscillatory. If the process is not oscillatory, then the range of the scale is divided into four areas: stationary, quasi-stationary, fractal and ultrafractal. The processes in the stationary region do not have a trend component. In the fractal region, processes are characterized by the presence of a trend component with a variable and constant trend. The ultrafractal region is occupied by processes having a nonlinear ultra-monotonically increasing (decreasing) trend. A quasi-stationary region combines the properties of a stationary and fractal regions; therefore, processes in this region have a hidden variable trend.

In Chap. 7 we shall continue to consider fractal and quasi-fractal scales and tie up them with probabilistic models of smart systems. The reasons for this are as follows:—According to [20] up to date, the dispute about the universality of the fractal nature of social processes remains unfinished. Specialists in whose area of interest this problem can be divided into two camps. The first includes those specialists who doubt the absolute fractality of the surrounding reality. The second camp includes those specialists who consider fractality to be a universal property, present to one degree or another in all synergetic nonlinear systems. They use following arguments. A fractal is a symbol of instability, clearly embodying the properties of the potential variety of alternative development options of the system. This is a structure that is a "balance" of order and chaos. From the position of synergetics, it is quite logical to present the structure of the evolutionary process as consisting of many such mutable, self-similar components. They propose to consider synergetics as a theory of evolving systems (which is consistent with the generally recognized scientific position), the mechanism of which is divided into three parts. The first and last are some unchanging limit states, the intermediate stage is a fractal, understood as a transition process. Other components of a similar system are also fractal. A closer look at the middle stage of evolution of a particular system always reveals that evolution goes through several stages of development before it unfolds from it. Moreover, each substage itself, in turn, is a fractal.

– In [1] it is noted that an important field of application of fractals is the analysis of time series: sequences of measuring values of indicators characterizing the phenomena or processes of quantities, and ordered in time.

As a rule, information on the behavior of complex systems is obtained in the form of precisely such experimental data. It is well known that fractals are graphs of realizations of extremely different processes, both stochastic (for example, Brownian motion) and deterministic (for example, the implementation of solutions of the logistic equation for certain parameter values).

In Chap. 7 we shall consider fractal and quasi-fractal scales.

References

1. Starchenko, N.V.: Fractality Index and Logical analysis of chaotic time series, 05.13.18—Mathematical modeling, numerical methods and program complexes 01.01.03—mathematical physics, dissertation for the degree of Ph.D. (physical and mathematical sciences) Moscow (2005) (in Russian)
2. Shcherbak, L.N.: On the Question of the Axioms of Measurement. National Aviation University, Kiev (2014). (in Russian)
3. Probability and mathematical statistics, Encyclopedia. Big Russian Encyclopedia, Moscow (1999), (in Russian)
4. Katulev, A.N., Severtsev, N.A.: Operations Research. In: Krasnoshchekov P (ed.) Decision making principles and security. Physical and mathematical literature, Moscow (2000) (in Russian)
5. Pfantsagl, I.: Theory of Measurements. Mir, Moscow (1976). (in Russian)

6. Serdyukova, N.A.: Optimization of Tax System of Russia, Parts I and II. Budget and Treasury Academy. Rostov State Economic University, Moscow (2002). (in Russian)
7. Home page. https://en.wikipedia.org/wiki/Level_of_measurement
8. Stevens, S.S.: On the theory of scales of measurement. Science **103**(2684), 677–680 (1946)
9. Mosteller, F.: Data Analysis and Regression: A Second Course in Statistics. Addison-Wesley Pub. Co., Boston, Reading, Mass (1977)
10. Chrisman, N.R.: Rethinking levels of measurement for cartography. Cartogr. Geogr. Inf. Sci. **25**(4), 231–242 (1998)
11. Orlov, A.I., Fedoseev, V.N.: Management in the technosphere. Publishing Center Academy, Moscow (2003)
12. Serdyukova, N., Serdyukov, V., Neustroev, S.: Testing as a Feedback in a Smart University and as a Component of the Identification of Smart Systems. In: Uskov, V., Howlett, R., Jain, L. (eds.) Smart Innovation, Systems and Technologies, SIST, vol. 144, pp. 527–538. Springer (2019)
13. Bavykin, O.B.: Fractal Multidimensional Scale, Designed to Control the Regime of Dimensional ECHO and Evaluate Its Output Data, Engineering Bulletin, FSBEI HPE MSTU named after N.E. Bauman (7), 559–566 (2013) (in Russian)
14. Mirkin, Y.M.: International practice of forecasting world prices in financial markets (raw materials, stocks, exchange rates), Institute of World Economy and International Relations of the Russian Academy of Sciences, Moscow (2014) (in Russian)
15. Malinetskii, G.G., Potapov, A.B.: Modern problems of nonlinear dynamics, URSS, Moscow (2000) (in Russian)
16. Gurevich, E.T., Prilepsky, I.V.: Analysis of expert and official forecasts of oil prices. Issues of Economics **4**, 26–48 (2018). (in Russian)
17. Wang, L., An, H., Xia, X., Liu, X., Sun, X., Huang, X.: Generating Moving Average Trading Rules on the Oil Futures Market with Genetic Algorithms, Hindawi Publishing Corporation Mathematical Problems in Engineering (2014), Home page http://dx.doi.org/10.1155/2014/101808
18. Danilov, V.R.: Genetic programming technology for generating automatic control systems for systems with complex behavior. Petersburg State University of Information Technologies, Mechanics and Optics, St. Petersburg (2007). (in Russian)
19. Kobenko, V., Yu.: Fractal Identification Scale, Omsk Scientific Bulletin (3), 205–213, (2009) (in Russian)
20. Maximova, M.V.: Fractals in modern synergetics, Izvestiya vuzov. North Caucasus region. Social Sciences **1**, 8–12 (2011). (in Russian)

Chapter 5
Testing Problems. Testing as a Coding of Knowledge System

Abstract In this chapter the following basic question is considered: The test system is a measurement system through assessments of students' knowledge. The main issues here are:

- the question of assessing the adequacy of the results of measuring the real level of knowledge and skills of students,
- the question of a comprehensive assessment of the level of assimilation of the system of knowledge that connects quantitative and qualitative indicators.

The main issues here are the question of assessing the adequacy of the measurement results to the real level of knowledge, skills of students. In this regard we shall consider two models of the algebraic formalization of the representation of the knowledge system—in the form of a free group of factors that determine the knowledge system, and in the form of a semantic network, which we represent using finite Boolean algebra. These models allow us to show that all binary tests i.e. tests involving answers in the form of either "yes" or "no", can be solved true, not knowing the specific nature of the proposed questions. Then we shall show that probability measure can be used as a measure of the level of assimilation of the knowledge system represented by the semantic network.

Keywords Testing · Binary tests · System efficiency · Knowledge system · Semantic network · Probability measure

5.1 Introduction. Main Measurement Problems in Test Field

Many researches claim that tests are a measure of the level of mastery of a knowledge system. However, there are no adequate enough quantitative assessments of the level of mastery of the knowledge system. Under rather strict restrictions, we shall construct such an estimate. It is shown that the tensor estimation of the system efficiency defined in [1, Chap. 6], including over a field of two elements, can be

© Springer Nature Switzerland AG 2021   83
N. A. Serdyukova and V. I. Serdyukov, *Algebraic Identification*
of Smart Systems, Intelligent Systems Reference Library 191,
https://doi.org/10.1007/978-3-030-54470-6_5

considered as encoding current state of a system. We shall construct a tensor estimate of the effectiveness of the functioning of the system as a homomorphism of a group of factors G, determining the system S into a group GL (n, R) of linear homogeneous transformations of the vector space R^n.

Let's list the main measurement problems in the field of testing, see also [2]:

1. Testing, like any of the currently known measurement methods, is not an absolutely objective universal method for measuring learning outcomes, since it depends on the specifics of knowledge, preferences, qualifications, etc. test developers. The levels of required knowledge of students determined by tests depend on these parameters. In addition, these measurements are not direct, but indirect, mediated through the results of the test.

2. In the measurement process, through testing, a limited number of characteristics that determine educational achievement are evaluated, and not their system, the knowledge of which can and should lead to synergistic effects in the learning process. In fact, these are synergistic effects that determine the quality of training. In this regard, the tests results provide very limited information on student training.

3. The problem of test validity is closely related with the problem 1. Validity is an indicator of the measure of compliance of a test to a test objective [3]. In the process of working out tests, the question in relation to the contradiction what one should to solve: a valid, but unreliable measurement, or not valid, but reliable measurement is always resolved. A test cannot be reliable if its validity is not ensured. The problem is what level of validity should be chosen.

4. Measurements always include measurement error. Testing is carried out, as a rule, at a time. If you conduct the same test (the same test version) twice, then the results will differ for different reasons. If the student performs another test option, the results may also be different. The inconsistency of the measurement results due to the different contents of the test or the conditions of the test determines the measurement error. This problem boils down to finding a way to determine the measurement error when using a specific test when developing a test.

5. Lack of communication with the measurement scales and the absence of a clearly defined unit. Testing Problems. Assume that the test consists of 30 tasks, arranged as their complexity increases. Let's consider two examples. Example 1: One student correctly completed 10 easy tasks, and the other—10 difficult ones, both got the same score. Is their background the same? Example 2: Three students correctly completed a different number of tasks: the first done 5 tasks, the second done 15 tasks, and the third done 25 tasks. Can one say that the preparation of the second student is also different from the preparation of the first, like the preparation of the third student differs from the preparation of the second? To none of these answers can be given a positive answer. This problem is solved by determining the scale of measurement and units of measurement when developing a test and a system for its evaluation.

5. The system of assessing the level of knowledge of students should not be inconsistent, and therefore the characteristics of the level of knowledge of students

measured in various ways should be interconnected. For example, test results should correlate with student grades. Thus, the characteristics of educational achievements are set in two ways, the first determines the content of the test, and the second determines how its results will be interpreted (connection with an external criterion).

6. The question of the correct assessment of the test results during testing should be attributed to the most important. The student completed some part of the tasks this is the virgin result. To translate the virgin result into a standardized indicator, one needs to have the data obtained on a representative sample from the population to which the subject belongs. To solve this problem, let's first show that testing can be considered as coding of a knowledge system. We show that testing is a coding of a knowledge system. Then we show that coding can be considered as a tool for measuring the complexity of a test. And after that we move on to solving problems 1 and 2.

5.2 Testing as a Coding of a Knowledge System

Algebraic formalization of smart systems, and, in particular, system identification matrices (in our case, knowledge systems and test systems) allow one to consider testing as a coding of a knowledge system [4].

In fact, let's consider alphabetical coding. Let $B = \{b_1, \ldots, b_m\}$ be an alphabet, B^* be a set of all words in the alphabet B, and $L \subseteq B^*$. Binary coding is a mapping $f : L \rightarrow \{0, 1\}^*$, where $\{0, 1\}^*$ is a set of all words in the binary alphabet $\{0, 1\}$. Herewith, the mapping f is an injective one. Let's designate by T the system of tests for the knowledge system L. Then $T \subseteq L$. Let $\varphi : T \rightarrow \{0, 1\}^*$ be a binary coding of a system of tests and $f : L \rightarrow \{0, 1\}^*$ be a binary coding of a knowledge system. Then mappings φ and f are injective ones.

Let's consider the diagram

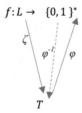

Then $\zeta = \varphi^{-1} f : L \rightarrow T$ is an injective mapping from L into T, that is coding of a knowledge system L with the help of system of tests T.

5.3 Coding as a Measure Students' Knowledge Tool

Further, here from we get that coding is a tool for measuring students' knowledge. That is, we get that coding in some cases (when considering a test system as a coding of a knowledge system) underlies the creation of measuring materials to assess the width and depth of students' knowledge. Let $\zeta = \varphi^{-1} f : L \to T$, on L and T the operation of composition of words is given, that is, their sequential writing out. So, we have semigroups $\langle L, \circ \rangle$ and $\langle T, \circ \rangle$, and we get the following diagram.

$$\langle T, \circ \rangle \xrightarrow{f_0} \langle \{0,1\}^*, \circ \rangle \xrightarrow{i} \langle Q, \circ, < \rangle \to \langle R, \circ, < \rangle$$

$$\langle \{0,1\}^*, \circ \rangle \to \langle \{0,1\}^*, \circ \rangle \to \langle \{0,1\}^*, \circ, < \rangle \subsetneqq \langle R, \circ, < \rangle$$

where i is a natural embedding $\{0, 1\}^*$ into the set of all rational numbers Q, obtained by assigning code 0 in front of each code and a comma, that is i is a mapping of all finite fractions of the form $0, a_1, a_2, \ldots, a_k$, where each $a_1, a_2, \ldots, a_k$ equals to either 0, or 1. Thus the order relation on the set $\{0, 1\}^*$ is defined.

Further, if E_n denotes the n-dimensional metric space of all binary vectors of the length n with the Hamming metric, then an arbitrary subset C of the space E_n is called a binary code of length n, the code elements are called codewords. Hamming distance is the number of positions in which the corresponding characters of two words of the same length are different[1].

A linear (or group) code is a subset of E_n, which is a linear subspace (subgroup) in E_n. Similarly, a linear q-valued code is a linear subspace of the n-dimensional metric space E_q^n of all vectors of the length n with the Hamming metric over the Galois field $GF(q)$, $q = p^k$, $q \geq 2$, where p is a prime number (such the code may not be a group code in E_q^n).

Suppose that the knowledge base B of the knowledge system S under consideration is written in the NPC language (Narrow Predicate Calculus) of signature Ω. If the elementary theory $Th(B)$ is decidable, then the set of tests $T \subseteq Th(B)$ is decidable that means that one can solve all the tests correctly without knowing the specific nature of the proposed questions. If $Th(B)$ is an unsolvable theory, then it is impossible to answer the tests correctly without knowing the essence of the questions. We shall consider now the representation of the knowledge system[2] L in the form of a free group G_L with a set $\{a_1, \ldots, a_m\}$ of generating elements, which are atomic

---

[1] Hamming distance: The number of digit positions in which the corresponding digits of two binary words of the same length are different.

[2] The representation of knowledge is a question that arises in cognitive science (the science of thinking), in computer science, and in the study of artificial intelligence. In cognitive science, it is related to how people store and process information. In computer science it is related with the selection of the presentation of specific and generalized knowledge, information and facts for the accumulation and processing of information in computers. The main task in artificial intelligence

elements of the knowledge system L. According to the theorem of Kharlampovich [6], on the solvability of the theory of free groups, it means that all binary tests can be solved correctly, without knowing the specific nature of the proposed questions, because the elementary theory $Th(G_L)$ is decidable.

Theorem 5.1 There exists an algorithm that allows one to solve all binary tests of the knowledge system S not knowing the specific essence of the proposed questions.

Let's concern the difference between encryption and encryption. Encryption is a way to modify a message or other document, which provides distortion (hiding) of its contents. Coding is the conversion of plain text into code. It is conceived that there is a one-to-one correspondence between the characters of the text (data, numbers, words) and the character code—this is the fundamental difference between coding and encryption. Formally, one can define encryption "in the form" of encoding, because encoding is a mapping of characters from one code to characters of another code, and, of course, the encryption algorithm also maps elements of a variety of plaintexts to elements of a variety of ciphertexts—this is where inaccuracies and errors grow. But, again, coding deals with the presentation of information in communication channels, in storage devices. Ciphers, encryption—this is a higher level [7].

Admittedly [7] earlier the terms "code" and "cipher", "coding" and "encryption" were used as synonyms. However, in modern conditions this is a mistake. What is the difference between code and cipher? When using any code, the message is first encoded on the transmitting side. The receiving side decodes this encoded message so that its true content to become clear. Similarly, a message is encrypted using a cipher, and then decrypted using the same cipher. A distinction between codes and ciphers exists. In cryptology, a code is a method used to encrypt a message that operates at the level of meaning; that is, words or phrases are converted into something else. The U.S. National Security Agency defined a code as "A substitution cryptosystem in which the plaintext elements are primarily words, phrases, or sentences, and the code equivalents (called "code groups") typically consist of letters or digits (or both) in otherwise meaningless combinations of identical length." [8, 9]. A codebook is needed to encrypt, and decrypt the phrases or words. By contrast, ciphers encrypt messages at the level of individual letters, or small groups of letters, or even, in modern ciphers, individual bits. Messages can be transformed first by a code, and then by a cipher. Such multiple encryption, or "super encryption" aims to make cryptanalysis more difficult. In cryptography, encryption is the process of encoding information [10, 11].

Codes are understood as methods and techniques of converting information using legend systems used to display and transmit certain information in a unique, but understandable and accessible way. At the same time, ciphers are methods and ways of converting information in order to protect it from illegal users. Both codes and ciphers are methods and techniques of converting information. Particular attention

---

(AI) is to learn how to store knowledge in such a way that programs can intelligently process it and achieve the likeness of human intelligence [5].

should be paid to why and for what purpose this conversion is carried out using codes and ciphers. The purpose of codes and ciphers is the main difference between them [10, 11].

5.4 Binary Tensor Estimation of System Efficiency as Encoding the Current State of the System

Now we shall show that a tensor estimate of the effectiveness of a system can be considered as encoding the system's current state. We shall consider a tensor estimate of the effectiveness of a system constructed over a field of two elements. According to [1] the tensor estimate of the system's efficiency, this is a mapping of the group of factors G_S, that determine the system S, into a complete linear group $GL(n, F)$ of the order n, i.e. into a group of all invertible matrices of order n over a field F of two elements, or, which is the same thing, into a group of invertible linear operators of the space F^n.

In [1, Chap. 6] the concept of tensor estimation of a system was introduced. Let's generalize this concept as follows.

Definition 5.2 A tensor estimate of the effectiveness of a system S over a field F is a mapping of the group of factors G_S, that determine the system S, into a complete linear group $GL(n, F)$ of the order n, that is the group of all invertible matrices of order n, or the group of invertible linear operators of the space F^n.

Definition 5.3 A binary tensor estimate of the effectiveness of a system S is its tensor estimate over a field of two elements Z_2, that is a mapping of the group of factors G_S, that determine the system S, into a complete linear group $GL(n, Z_2)$ of the order n, that is the group of all invertible matrices of the order n, or the group of invertible linear operators of the space Z_2^n.

So, one obtains from Definition 5.3 that the binary tensor estimate of the effectiveness of the system S is the encoding of the current state of the system S.

We represent the tensor estimate as a homomorphism $G_{S_i} \to GL(n, R)$, where G_{S_i} is a group of factors defining the system S_i, $GL(n, R)$ is a group of linear homogeneous transformations of a vector space R^n, n—the number of quantitative indicators that assess the quality of the subsystem G_{S_i} of the system S.

In [1, Chap. 6] we decomposed the knowledge system into:

1. Subsystem of knowledge S_1 (an informational subsystem of the system S)
2. Subsystem of pedagogy or methodological and methodical subsystem S_2 (an adaptive subsystem of the system S)
3. Subsystem of students S_3 (a target subsystem of the system S)
4. A financial subsystem S_4 (a providing subsystem of a system S).

According to this decomposition one can single out the following goals which are correspond to the distinguished subsystems:

- The purpose of the informational system S_1 is to present (have) all the necessary information of the knowledge system S.
- The purpose of the adaptive system S_2 is to present all the necessary methodological and methodological support of the knowledge system S, adapted for the target audience S_3.
- The purpose of the system of the target audience (a target subsystem) S_3 is to assimilate (study) the informational system S_1.
- The purpose of the supporting (financial) system S_4 is to provide the necessary financial resources to ensure the functioning of the system S.

In turn, the system for assessing learning outcomes is divided into subsystems:

- assessments of individual educational achievements of students (state or growth dynamics) for certification purposes (confirmation obtaining a certain level of education), correction of individual students' results, transition to the next level of training, the choice of the level of study of individual disciplines;
- assessments of the level of educational achievements of a group, university, educational institution with the aim of evaluating the activities of the teaching staff or (or), improving the teaching and learning process;
- monitoring the educational achievements of a sample of students across the regions or the country as a whole in order to assess the quality of education and development trends.

Thus, in determining the theoretical and methodological approaches to the construction of subsystems for assessing learning outcomes, the following questions play a decisive role:

1. For what purpose will the system being created be used? What will be evaluated?
2. How objective, reliable and valid should the results should be? Is their interpretation possible?
3. What conclusions should be made based on these results?
4. What problems can arise in the process of developing tools and conducting an assessment?

We have already shown in Sect. 5.2 that testing is a coding of a knowledge system, that is, testing can be interpreted as coding a knowledge system. Let's show now that coding can be considered as a tool for measuring the level of students' knowledge. Then we move on to solving problems 1 and 2 formulated in Sect. 5.1

The description of the abstract system using models of factors determining the system makes it possible to construct very diverse models of the studied system S. In this chapter, we consider the representations of the knowledge system, and, in particular, the model G_L of the knowledge system L c with the help of a free group with a set $\{a_1, \ldots, a_m\}$ of generating elements, which are atomic elements of the knowledge system L, and semantic networks.

In this chapter we shall consider the following issue: how are related models of factors that determine the knowledge system and the model of the semantic network of the knowledge system.

According to [13], a semantic network is an information model of a subject area, which has the form of a directed graph, whose vertices correspond to objects of the subject area, and edges define the relationship between them. Objects of a subject area can be concepts, events, properties, processes. Thus, the semantic network is one of the ways of representing knowledge. The title combines terms from two sciences: semantics in linguistics studies the meaning of language units, and the network in mathematics is a kind of graph—a set of vertices connected by edges to which a certain number is usually assigned. The role of vertices is played by the concepts of a knowledge base, and edges (moreover, directed ones) define relations in a semantic network between the concepts of a knowledge base. Thus, the semantic network reflects the semantics of the subject area in the form of concepts and relations.

5.5   Representation of the Semantic Model of the Knowledge System (Semantic Network of the Knowledge System) in the Form of a Finite Boolean Algebra

Let's imagine a semantic model of a knowledge system in the form of a finite Boolean algebra $B = \langle \langle \{F_1, \ldots, F_n\}, \rangle \wedge, \vee, \rceil, 0, 1 \rangle$, where $\wedge$ is a conjunction, $\vee$ is a disjunction, $\rceil$ is a negation, 0 is "false", 1 is "true". By $\langle \{F_1, \ldots, F_n\} \rangle$ we shall designate the closure of the set of atomic elements of the semantic network of the knowledge system with respect to the operations $\wedge, \vee, \rceil, 0, 1$. Then B is a finite Boolean algebra. Each Boolean algebra is a distributive lattice with complements, and, conversely, each distributive lattice with complements, is a Boolean algebra [14]. The elementary theory $Th(B)$ describes the knowledge system S.

We use the corollary from [15–17], **Stone's theorem**:

Any two finite Boolean algebras of the same power are isomorphic.

Definition 5.4 An algebra $B = \langle \langle \{F_1, \ldots, F_n\} \rangle, \wedge, \vee, \rceil, 0, 1 \rangle$, where $\wedge$ is a conjunction, $\vee$ is a disjunction, $\rceil$ is a negation, 0 is "false", 1 is "true", is called the closure of the semantic network $\{\{F_1, \ldots, F_n\}, \wedge, \vee, \rceil, 0, 1\}$.

A finite Boolean algebra by Stone's theorem is an algebra of sets; therefore, there are 2^n in it for some natural n. Any finite Boolean algebra is atomic. If it has 2^n elements, then n is the number of its atoms. We now consider the finite Boolean algebra as a finite distributive lattice with complements. Since the finite set algebra is a σ—algebra, then one can define a probability measure on B.

In [16, p. 43], the following theorem is proved:

Theorem 5.5 [16] The Stone space of a Boolean algebra A is metrizable if and only if A is no more than countable.

Now let's give examples from [16] illustrating these concepts.

(1) The ring Z_2 of all residue classes of integers modulo 2, considered as a Boolean algebra, is a two-element Boolean algebra.
(2) The ring Z_2^X regarded as a Boolean algebra, is isomorphic to the field of all subsets of the set X.

An atom of Boolean algebra is its minimal nonzero element.

In [17], it was proved that the elementary theory of a finite Boolean algebra is decidable; therefore, we get another, simpler proof of Theorem 5.1.

Consider the following examples from [14, 18]

Example 5.6 Any two dense linear orders without endpoints are elementarily equivalent. Any two atomless Boolean algebras are elementarily equivalent. An elementary classification of the class K of algebraic systems is a criterion that describes when two K-systems are elementarily equivalent. For finite σ-systems A and B, the relations of isomorphism and elementary equivalence are equivalent. In [18] also shown that the following theorem takes place.

Theorem [18] A sentence Φ is stable with respect to (a) subsystems, (b) extensions, (c) the union of upward systems, (d) homomorphic images if and only if Φ is equivalent to some (a) $\forall$-sentence, (b) $\exists$-sentence, (c) $\forall\exists$-sentence, (d) a positive or identically false sentence, respectively.

The following Birkhoff theorem is a well-known one.

Theorem, G. Birkhoff [14, 18]. A sentence Φ is equivalent to some identity if and only if Φ is stable with respect to subsystems, homomorphic images, and direct products.

For finite algebraic systems, the relations of their isomorphism and elementary equivalence are equivalent [17, 19].

It is well-known that any finitely generated subalgebra of Boolean algebra is finite, any finite Boolean algebra is generated by the set of all atoms.

5.5.1 Probabilities on Semantic Networks. Probability Measure as a Measure of the Level of Assimilation of the Knowledge System Represented by the Semantic Network

On σ—algebra, which is the set of all subsets of a finite set Ω, the probability is completely determined if one sets it for all events consisting of one elementary outcome. That can be done in the following way. Let $\omega = (\varepsilon_1, \varepsilon_2, \ldots, \varepsilon_n)$—a set of zeros and ones, we assume the probability

$$p(\{\omega\}) = q^{\sum_{i=1}^{n} \varepsilon_i}(1-q)^{n-\sum_{i=1}^{n} \varepsilon_i} \tag{5.1}$$

Here experiment is a choice the answer to the test F_i. Event A consists in choosing the right answer to the test. Let the probability of choosing the correct answer to the test $F_i, i = 1, \ldots, n$, be independent of i and be equal to q.

Formula (5.1) takes place if one has the Bernoulli test scheme, that is, the experiments are carried out under constant conditions and the experiments are independent from each other. In this case, the experiments are independent in aggregate and we have Bernoulli scheme. If X is the number of occurrences of event A in n trials, then with $n \to \infty$ we have $\frac{X}{n} \to q$, that is the average $\frac{X}{n}$ equals q.

This shows that by varying q through promoting the training, one can achieve sufficiently high testing results. This means that with the help of q we can measure the level of assimilation of the knowledge system, that is, q is a measure of the level of assimilation of the knowledge system.

On the other hand, theorem 5.1 shows that there is an algorithm which allows to solve all binary tests of the knowledge system without knowing the specific nature of the proposed questions.

Let now the probability of choosing the right answer to the test $F_i, i = 1, \ldots, n$, equals $q_i, i = 1, \ldots, n$, where $F_i, i = 1, \ldots, n$, is a set of binary tests,$\omega = (\varepsilon_1, \varepsilon_2, \ldots, \varepsilon_n)$—a set of zeros and ones, showing whether the answer is true or false for each test. Then $p(\{\omega\}) = \prod_{i=1}^{n} \alpha_i$, where

$$\alpha_i = \begin{cases} q_i, \, if \, \varepsilon_i = 1 \\ 1 - q_i, \, if \, \varepsilon_i = 0 \end{cases}$$

Let's now extend the function p to the entire σ—algebra. This is possible, since on the σ—algebra which is the set of all subsets of a finite set Ω, the probability is completely determined if you set it for all events consisting of one elementary outcome. On the closure $\boldsymbol{B} = \langle\langle\{F_1, \ldots, F_n\}\rangle, \wedge, \vee, \rceil, 0, 1\rangle$ of the semantic network $\{\{F_1, \ldots, F_n\}, \wedge, \vee, \rceil, 0, 1\}$ let's define a probability measure $p : \langle\{F_1, \ldots, F_n\}\rangle \to [0, 1]$, and then consider the restriction $p\lceil\{F_1, \ldots, F_n\}$.

Then the probability measure p is a measure of the level of assimilation of the knowledge system represented by the semantic network $\{\{F_1, \ldots, F_n\}, \wedge, \vee, \rceil, 0, 1\}$.

5.5.2 Homomorphisms that Define Measures

Next on the closure $\boldsymbol{B} = \langle\langle\{F_1, \ldots, F_n\}\rangle, \wedge, \vee, \rceil, 0, 1\rangle$ of the semantic network $\{\{F_1, \ldots, F_n\}, \wedge, \vee, \rceil, 0, 1\}$ let's define a probability measure $p : \langle\{F_1, \ldots, F_n\}\rangle \to [0, 1]$, and then consider the restriction $p\lceil\{F_1, \ldots, F_n\}$. As $F_1, \ldots, F_n$ are atoms of the semantic network then $\boldsymbol{B} = \langle\langle\{F_1, \ldots, F_n\}\rangle, \wedge, \vee, \rceil, 0, 1\rangle$ is finite by Stone's theorem, consists of 2^n elements and is isomorphic to the Boolean algebra of sets $2^{\{F_1,\ldots,F_n\}}$.

According to [20], p is a normalized measure. Let $\mathbf{F} = \langle F, \cup, \cap, \backslash,'\rangle$ be a finite Boolean algebra. Let's consider $\mu : \mathbf{F} \to R^+ \cup \{\infty\} \cup \{0\}$ or $\mu : \mathbf{F} \to R^+ \cup \{\infty\} \cup \{0\}/\mathbf{Z}^+ \cup \{\infty\} \cup \{0\}$ such that $\mu(A \cup B) = \mu(A) + \mu(B)$, then μ is a homomorphism from semigroup $\langle F, \cup\rangle$ into a semigroup $\langle R^+ \cup \{\infty\} \cup \{0\}, +\rangle$. Now let's single out those homomorphisms μ, that are measures. For the probability measure μ of a space $2^{\{F_1,\ldots,F_n\}}$ one has:

$$\mu(A \cup B) = \mu(A) + \mu(B), if\, A \cap B = \emptyset, A, B \in \langle\{F_1, \ldots, F_n\}\rangle$$

For homomorphism $\mu :< 2^{\{F_1,\ldots,F_n\}}, \cup >\to \langle R^+ \cup \{\infty\} \cup \{0\}, +\rangle$ or $\mu : \mathbf{F} \to R^+ \cup \{\infty\} \cup \{0\}/\mathbf{Z}^+ \cup \{\infty\} \cup \{0\}$ one has:

$$\mu(A \cup B) = \mu(A) + \mu(B), \forall A, B \in 2^{\{F_1,\ldots,F_n\}}.$$

$$\bigcup_{i=1}^{\infty} A_i = \lim_{\to}\left\{\bigcup_{i=1}^{n} A_{i|n=1,2,\ldots}\right\}$$

(here we presented an infinite union of elements of closure of the semantic network, standing on the left side of the equality in the form of a direct limit of finite associations of elements of closure of the semantic network)

$$\mu\left(\bigcup_{i=1}^{\infty} A_i\right) = \mu\left(\lim_{\to}\left\{\bigcup_{i=1}^{n} A_i | i = 1, 2, \ldots, n\right\}\right).$$

applied to both sides of the equality μ.

If $\mu\left(\lim_{\to}\left\{\bigcup_{i=1}^{n} A_i | i = 1, 2, \ldots, n\right\}\right) = \lim_{n\to\infty}\sum_{n=1}^{n}\mu(A_i)$, then we have

$$\mu\left(\bigcup_{i=1}^{\infty} A_i\right) = \mu\left(\lim_{\to}\left\{\bigcup_{i=1}^{n} A_i | i = 1, 2, \ldots, n\right\}\right) = \lim_{n\to\infty}\sum_{n=1}^{n}\mu(A_i) = \sum_{i=1}^{\infty}\mu(A_i)$$

The last equality always holds, since the closure of the semantic network $\mathbf{B} = \langle\langle\{F_1, \ldots, F_n\}\rangle, \wedge, \vee, \urcorner, 0, 1\rangle$ is a finite Boolean algebra of 2^n elements.

Theorem 5.7 There exists a measure μ on the closure of a finite semantic network $\mathbf{B} = \langle\langle\{F_1, \ldots, F_n\}\rangle, \wedge, \vee, \urcorner, 0, 1\rangle$ which is not a homomorphism of the finite semigroup $\langle F, \cup\rangle$ into the semigroup $R^+ \cup \{\infty\} \cup \{0\}/\mathbf{Z}^+ \cup \{\infty\} \cup \{0\}$..

The proof follows from the fact that $\mu(A \cup B) \neq \mu(A) + \mu(B)$, if $A \cap B \neq \emptyset$.

Theorem 5.8 Any homomorphism $\mu : \mathbf{F} \to R^+ \cup \{\infty\} \cup \{0\}/\mathbf{Z}^+ \cup \{\infty\} \cup \{0\}$ of the closure of a finite semantic network is a measure on the closure of a finite semantic network.

Remark Normalizing μ, we get a probability measure, $p = \frac{1}{\mu(\Omega)}\mu$, $\Omega = \{F_1, \ldots, F_n\}$—is the set of all elementary events of Boolean algebra $2^{\{F_1, \ldots, F_n\}}$.

5.5.3　Quasi-fractal Measure as a Quasi-fractal Homomorphism

Let's now generalize the notion of measure up to the notion of quasi-fractal measure, using the notion of a quasi-fractal homomorphism, Chap. 3, Sect. 3.3, Definition 3.2. as the analogue of the notion that we should to construct. Also, we need the general definition of a measure, see, for example [21].

Definition [21] A measure on a set X is an arbitrary function $\mu : K \rightarrow R^+$, defined on some ring of sets K on X, and satisfying the additivity condition: $(\forall A, B \in K)(A \cap B = \emptyset \Rightarrow \mu(A \cup B) = \mu(A) + \mu(B))$.

An ordered triple $\langle X, K, \mu \rangle$, where $\mu : K \rightarrow R^+$ is a measure, is called a measure space; sets from K are called μ-measurable. A space with measure can also be considered a pair $\langle X, \mu \rangle$, assuming μ-measurable sets to be already defined and giving a ring of sets on X. Any measure $\mu : K \rightarrow R^+$ on X has the following properties:

1. $\mu(\emptyset) = 0$.
2. $(\forall A, B \in K)(A \subseteq B \Rightarrow \mu(A) \leq \mu(B))$ (monotony of measure).
3. If K an algebra of sets on X, then $\mu(X)$ is the largest value of the measure μ.

So, we get from here the following definition of a quasi-fractal measure.

Definition 5.9 Let $A_1 = \langle A_1; \Omega_1 \rangle$ be a quasi-fractal algebraic system. A quasi-fractal measure on a quasi-fractal system $A_1 = \langle A_1; \Omega_1 \rangle$ is a quasi-fractal function
$\mu : K = \langle K, \cup, \cap, \backslash \rangle = \langle A_1; \Omega_1 \rangle \rightarrow R^+ = \langle A_1; \Omega_1 \rangle$ defined on some quasi-fractal ring of sets $K = \langle K, \cup, \cap, \backslash \rangle = \langle A_1; \Omega_1 \rangle$ on X_k, k is a level of a quasi-fractal ring $K = \langle K, \cup, \cap, \backslash \rangle = \langle A_1; \Omega_1 \rangle$ and satisfying the additivity condition on each level k:

$$\left(\forall A, B \in K_\mu\right)(A \cap B = \emptyset \Rightarrow \mu_k(A \cup B) = \mu_k(A) + \mu_k(B)).$$

An ordered quasi-fractal triple $\langle X, K, \mu \rangle = \langle A_1; \Omega_1 \rangle$, where $\mu : K \rightarrow R^+$ is a quasi-fractal measure, is called a measure quasi-fractal space; quasi-fractal sets from K are called μ-measurable. A quasi-fractal space with measure can also be considered a pair $\langle X, \mu \rangle = \langle A_1; \Omega_1 \rangle$, assuming μ_k-measurable sets to be already defined and giving a ring of sets on X_k. Any measure $\mu_k : K_k :\rightarrow R^+$ on X has the following properties:

1. $\mu(\emptyset) = 0$.

2. $(\forall A, B \in K_k)(A \subseteq B \Rightarrow \mu_k(A) \leq \mu_k(B))$ (monotony of measure).
3. If K_k an algebra of sets on X_k, then $\mu_k(X_k)$ is the largest value of the measure μ_k.

Theorem 5.10 There exists a quasi-fractal measure μ on the closure of a quasi-fractal finite semantic network $B = \langle\langle\{F_1, \ldots, F_n\}\rangle, \wedge, \vee, \rceil, 0, 1\rangle = \langle A_1; \Omega_1 \rangle$ which is not a quasi-fractal homomorphism of the finite quasi-fractal semigroup $\langle F, \cup \rangle = \langle A_1; \Omega_1 \rangle$ into the semigroup $R^+ \cup \{\infty\} \cup \{0\}/Z^+ \cup \{\infty\} \cup \{0\} = \langle A_1; \Omega_1 \rangle$.

The proof follows from the fact that for each level k $\mu_k(A \cup B) \neq \mu_k(A) + \mu_k(B)$, if $A \cap B \neq \emptyset$.

Theorem 5.11 Any quasi-fractal homomorphism $\mu : F = \langle A_1; \Omega_1 \rangle \rightarrow R^+ \cup \{\infty\} \cup \{0\}/Z^+ \cup \{\infty\} \cup \{0\}$ of the quasi-fractal closure of a quasi-fractal finite semantic network is a measure on the closure of a quasi-fractal finite semantic network.

Remark Normalizing μ_k on each level of a quasi-fractal, we get a probability measure, $p_k = \frac{1}{\mu(\Omega)}\mu_k$, $\Omega_k = \{F_{k1}, \ldots, F_{kn}\}$—is the set of all elementary events of Boolean algebra $2^{\{F_{k1}, \ldots, F_{kn}\}}$.

To prove these theorems, it is enough to remained the Definition 3.2 of a concept of a quasi-fractal homomorphism from Chap. 3:

Let $A_1 = \langle A_1; \Omega_1 \rangle$, $B_1 = \langle B_1; \Omega_1 \rangle$ be quasi-fractal algebraic systems. A mapping $f : A_1 = \langle A_1; \Omega_1 \rangle \rightarrow B_1 = \langle B_1; \Omega_1 \rangle$ is called a quasi-fractal homomorphism from a system $A_1 = \langle A_1; \Omega_1 \rangle$ into a system $B_1 = \langle B_1; \Omega_1 \rangle$, if the following two conditions are fulfilled:

(1) For every $\alpha \in \Lambda_k$ the equality $f(a_\alpha) = f(b_\alpha)$ holds for some $b_\alpha \in \Lambda_k$, that is f maps the elements of the quasi-fractal algebraic system $A_1 = \langle A_1; \Omega_1 \rangle$ of the level $k + 1$ into elements of the same level $k + 1$ of the quasi-fractal algebraic system $B_1 = \langle B_1; \Omega_1 \rangle$;
(2) f saves all operations and predicates of the signature Ω_k for all levels k of the quasi-fractal.

References

1. Serdyukova, N., Serdyukov, V.: Algebraic Formalization of Smart Systems. Theory and Practice, Smart Innovation, Systems and Technologies, SIST, vol. 91, Springer Nature, Switzerland (2018)
2. Program for the Development and Improvement of State Educational Standards and Testing (First Stage) Final Report. Appendix 4. Foreign construction experience and current problems of the development of educational testing
3. Home page. https://psychology.academic.ru/3954
4. Solovieva, F.I.: Introduction to Coding Theory. Novosibirsk State University, Novosibirsk (2006)
5. Home page. https://en.wikipedia.org/wiki/Cognitive_science

6. Kharlampovich, O., Myasnikov, A.: Elementary theory of free non-abelian groups. J. Algebra **302**(2), 451–552 (2006)
7. Adamenko, M.: Fundamentals of classic cryptology. Mashinostroenie publ., Moscow, Secrets of ciphers and codes (2014), Home page. http://wmhelp.net/lib/b/book/3167518757/19
8. Boak, D.G.: A History of U.S. Communications Security; the David G. Boak Lectures National Security Agency (NSA), (I), (1973), (II), 1981, partially released (2008), additional portions (2015) Home page. http://www.governmentattic.org/18docs/Hist_US_COMSEC_Boak_NSA_1973u.p
9. Home page. https://en.wikipedia.org/wiki/Code
10. Agrawal, M.: A comparative survey on symmetric key encryption techniques. Int. J. Comput. Sci. Eng. **4**, 877–882 (2015)
11. Home page. https://en.wikipedia.org/wiki/Encryption
12. Home page. https://en.wikipedia.org/wiki/Knowledge_representation_and_reasoning
13. Home page. https://en.wikipedia.org/wiki/Semantic_network
14. Malt'sev, A.I.: Algebraic Systems. Nauka, Moscow (1970) (in Russian)
15. Sultanbekov, F.F.: From Lattices to Boolean Algebras. Kazan (Volga) Federal University, Kazan (2012). (in Russian)
16. Sikorski, R.: Boolean algebra, Mir, Moscow (1969) (in Russian)
17. Goncharov, S.S.: Countable Boolean algebras and solvability. Scientific book, Siberian School of Algebra and Logic, Novosibirsk (1996)
18. Bunina, E.I., Mikhalev, A.V., Pinus, A.G.: Elementary and close to it logical equivalences of classical and universal algebras. Publishing House MTsNMO, Moscow (2015)
19. Gurov, S.I.: Ordered sets and universal algebra (introductory course), study guide. Moscow State University named after M.V. Lomonosov, Faculty of Computational Mathematics and Cybernetics, Moscow (2005)
20. Chernova, N.I.: Probability Theory. Novosibirsk State University, Novosibirsk (2007)
21. Vechtomov, E.M.: Basic Concepts of Algebra. Publishing house Raduga-PRESS, Kirov (2013)

Chapter 6
Smart System's Potential

Abstract In this chapter, we introduce the concept of parametric algebraic potential in line with the algebraic formalization of the system.

Keywords Potential · Algebraic formalization of the system

6.1 Introduction

In various fields of knowledge related to the humanities, such as, for example, economics, teaching theory, control theory, analogues of physical concepts are used, and, in particular, the concept of potential.

In [1] it is marked the following. "The potential theory has been studied very much, especially after the researches of Gauss in 1840, where he studied important problems and methods which gave yet remained partly as basic ideas of modern researches in this field. For about thirty years many refinements of the classical theory were given; later the axiomatic treatments starting from different particular aspects of the classical theory. About half a dozen of such axiomatic approaches to potential theory, parts of which are not yet published with details, exist". According to [2], Potential theory arose originally as a part of celestial mechanics studying the properties of gravitational forces acting according to the law of universal gravitation. The main results in the creation and initial development of the Potential theory belong Newton, Lagrange, Legendre, Laplace. Thus, Lagrange showed that the field of gravitational forces is potential. Beginning from Gauss' works, the notion of potential is used also in problems of electrostatics and magnetism, and "masses" (charges, magnetization) of an arbitrary sign began to be considered as potentials. In the nineteenth century, the main boundary-value problems were distinguished: the Dirichlet problem, the Neumann problem, the Robin problem, the mass sweeping problem (balayage method). Lyapunov and Steklov made a significant contribution to the study of basic boundary-value problems at the end of the nineteenth century. The results of the theory were essentially generalized at the beginning of the twentieth

© Springer Nature Switzerland AG 2021　　　　　　　　　　　　　　　97
N. A. Serdyukova and V. I. Serdyukov, *Algebraic Identification of Smart Systems*, Intelligent Systems Reference Library 191,
https://doi.org/10.1007/978-3-030-54470-6_6

century using the apparatus of measure theory and generalized functions. Subsequently, the theory of potentials involved analytical, harmonic and subharmonic functions, tools of probability theory.

In the 1950s, based on the methods of topology and functional analysis, an axiomatic abstract theory of potentials was developed.

In this chapter, we introduce the concept of parametric algebraic potential in line with the algebraic formalization of the system.

Let's remind that the potential energy [1] (in our context is a measure[2] of the transition of motion and interaction of systems or chaos) of a system from one form to another is determined by the mutual position of interacting bodies or parts of the same body, that is, of the structure of the system.

Potential energy is a scalar physical quantity, which is a part of the total mechanical energy of the system located in the field of conservative forces. It depends on the position of the material points that make up the system, and characterizes the work performed by the field when moving them.[3]

In mathematics, the concepts of measure, measurability, and integrability are closely related. The initial concept is a measure as a non-negative real function defined on some ring of subsets of one or another set.

Definition 6.1 [6] A measure on a set X is an arbitrary function $\mu : K \rightarrow R^+$, defined on some ring of sets K on X, and satisfying the additivity condition:

$$(\forall A, B \in K)(A \cap B = \emptyset \Rightarrow \mu(A \cup B) = \mu(A) + \mu(B)).$$

An ordered triple $\langle X, K, \mu \rangle$, where $\mu : K \rightarrow R^+$ is a measure, is called a measure space; sets from K are called μ-measurable. A space with measure can also be considered a pair $\langle X, \mu \rangle$, assuming μ-measurable sets to be already defined and giving a ring of sets on X. Any measure $\mu : K \rightarrow R^+$ on X has the following properties:

1. $\mu(\emptyset) = 0$.
2. $(\forall A, B \in K)(A \subseteq B \Rightarrow \mu(A) \leq \mu(B))$ (monotony of measure).
3. If K is an algebra of sets on X, then $\mu(X)$ is the largest value of the measure μ.

Let's recall the definition of a ring of sets.

---

[1]Energy (dr. Greek ἐνέργεια-action, activity, strength, power) is a scalar physical quantity, which is a single measure of various forms of motion and interaction of matter, a measure of the transition of matter from one form to another. The introduction of the concept of energy is convenient in the case if the physical system is closed, then its energy is stored in this system for the time during which the system will be closed This statement is called the law of conservation of energy. From a fundamental point of view, energy is one of the three (energy, momentum, angular momentum) additive integrals of motion (that is, quantities that remain during motion), related, according to Noether's theorem, to the homogeneity of time [3].

[2]Measure is a philosophical category expressing the organic unity of the qualitative and quantitative certainty of an object or phenomenon [4].

[3]Ref. [5].

Definition 6.2 A ring of sets on a set X is any nonempty set K of subsets in X, that is closed under set-theoretic operations of union $\cup$, intersection $\cap$ and difference:

$$(\forall A, B \in K)(A \cup B \in K) \wedge (A \cap B \in K) \wedge (A \backslash B \in K)$$

Any ring of sets K on a set X contains an empty set $\emptyset$ and a symmetric difference: $(\forall A, B \in K)(A \oplus B = (A \backslash B) \cup (B \backslash A) \in K)$ and K is an abstract ring with respect to the operations of addition $\oplus$ and multiplication $\cap$. A ring of sets K on a set X is called an algebra of sets on X, if $X \in K$. Every algebra of sets K on X together with each of its element A contains its complement $X \backslash A$, i.e., it is a Boolean algebra—a subalgebra of a Boolean $B(X)$.

Let's remained some well-known definitions:

- Borel function is a mapping of one topological space to another (usually both of them are the space of real numbers) for which the inverse image of any Borel set is a Borel set.
- Borel measure is a measure defined on all open (and, therefore, on all Borel) sets of a topological space.

Definition 6.3 By the algebraization of potential energy or the potential measure of the transition of motion and interaction or the potential property of a system, we call the possibilities determined by the (algebraic) structure of the system.

Kinetic energy always characterizes the body relative to the selected reference frame, that is, it is determined by the strength of the system's links (from the point of view of the algebraic formalization of systems—by the levels of links). The higher the level, the weaker is the connection. The kinetic energy of the body is determined by its speed relative to the selected reference frame; potential is determined by the location of bodies in the field. In a closed system, the total energy is equal to the sum of the potential and kinetic energies.

Definition 6.4 By the algebraization of kinetic energy, or the kinetic measure of the transition of motion and interaction, or the kinetic property of a system, we call the strength of the system's bonds, determined by its (algebraic) structure.

6.2 Main Construction

6.2.1 The Concept of System Potential

We need the following definitions from [7].

Definition 6.5 Let S be a system and G_S be a group of factors that determined the system S. The measure $PC(G_S)$ of the system S links strength is the number of possible different synergetic effects of the system S, that is the number of possible different final states of the system S, which are calculated by the model G_S, or, which is the same, the number of pairwise nonisomorphic groups of order $|G_S|$.

Notes

(1) Than $PC(G_S)$ index is larger, that the system S links calculated on the model G_S is weaker.
(2) Than $PC(G_S)$ index is smaller, that the system S links calculated on the model G_S is stronger.
(3) The system links indicator $PC(G_S)$ is a relative one and it depends on the choice of the model G_S of factors that determined the system S.
(4) Let's consider $pc(G_S) = \frac{1}{PC(G_S)}$.

Than the larger $pc(G_S)$ are the stronger the binding forces of the system S, calculated by the model G_S are. The smaller are $pc(G_S)$, then the weaker the binding forces of the system S, calculated by the model G_S are.

Definition 6.6 Let S be a system and G_S be a group of factors that determined the system S. Let $\emptyset \neq M \subseteq G$. The measure (M) of the set M links strength is the number of possible different synergetic effects of the system $\langle G\backslash M \rangle$, where $\langle G\backslash M \rangle$ is a subgroup of the group G_S, generated by the set $G\backslash M$.

Notes

(1) Than $PC(M)$ index is larger, that the set M links calculated on the model G_S is weaker.
(2) Than $PC(M)$ index is smaller, that the set M links calculated on the model G_S is stronger.
(3) The system links indicator $PC(M)$ is a relative one and it depends on the choice of the model G_S of factors that determined the system S.
(4) Let's consider $pc(M) = \frac{1}{PC(M)}$. Then the larger is $pc(M)$, the stronger are the binding forces of the set M of the system S, calculated by the model G_S. The smaller is $pc(G_S)$, the weaker are the binding forces of the set M of the system S, calculated according to the model G_S.

Since the system in algebraic formalization is determined by the factors that determine the system, and they can be different, and the models of the system, respectively, can differ, we shall talk about the parametric potential and parametric kinetic energy. Potential energy always characterizes the body relative to the source of force (force field). In [8], it is noted that existing methods of mathematical modeling reflect either processes (metric spaces of set-theoretic topology), or the structure of systems (combinatorial topology). One of the first to play the role of structure in technical systems was Gabriel Cron [9]. He proposed a model and equations for a generalized electric machine. All electric machines were considered as changes in the connections of elements of a generalized machine, i.e. differed in structure [10]. Then a tensor analysis of networks was developed [11] and a method for modeling and studying complex systems in parts—Diakoptics [12].

Now let's touch the notion of potential. According to [13], the notion "potential" an be characterized as the totality of all available capabilities, funds in any field, sphere, in the broad sense, "spare" funds. Consider now the definition of Newton's potential in three-dimensional Euclidean space R^3, that is

$$\int\limits_C \frac{d\mu(y)}{|x - y|}$$

where μ is continuous function on $C \subseteq R^3$.

The Newton's potential of a simple layer in integral form is represented by the formula:

$$V^{(0)}(x) = \iint\limits_C \frac{\mu(y)}{|x - y|} dC_y$$

where μ is a continuous function on a bounded piecewise smooth surface $C \subseteq R^3$.

To use the concept of potential, we shall pull on the surface onto the structure (lattice) of subgroups of the group G_S of factors determining the system S. So:

1. Let's construct a flat graph of the lattice of subgroups of the group G_S. We arrange ordered pairs at the vertices of the graph (x_V, y_V), where x_V is a reciprocal of the number of synergistic effects of a subgroup of a group G_V, corresponding to vertex V, that is $pc(G_V) = \frac{1}{PC(G_V)}$, and y_V is a communication level i.e. maximum strength between G_V and subgroups of the group G_S, incident to the vertex V in the lattice of subgroups of the group G_S, which is defined as follows (Fig. 6.1):
2. We now assume equal in Definition 6.6 one by one $M = G_V \cup G_{V_i}$, where $i = 1, \ldots, k$, and select the maximum value $pc(M) = \frac{1}{PC(M)}$.
    $x = pc(G_V)$, $y = pc(M)$, where $M = G_V \cup G_{V_i}, i = 1, \ldots, k$.
    The difference $x - y$ is the analog of a distance between x и y.
    Items (1) and (2) may be fulfilled in accordance with the Definition 2.2, Chap. 2.

Fig. 6.1 A part of the lattice
of subgroups of the group G_S

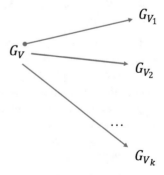

Definition 6.7 Using the Lagrange theorem, we pull on the surface (curve) μ in the domain C to the points $\{ (x_V, y_V) | V$ runs over the set of vertices of the lattice graph of subgroups of the group $G_S \}$ and consider the integral

$$\int_C \frac{d\mu(y)}{|x - y|}$$

The integral $\int_C \frac{d\mu(y)}{|x-y|}$ will be called the algebraization of the potential of the system S according to the model G_S.

The question about the possibility and necessity of integration on graphs are discussed in many works. In [14], for example, the integration by parts formula is extended for an integral in which the integration is carried out along the geometric graph. The need for integration by parts arises, for example, in the modeling of deformations and oscillatory processes of objects located along a geometric graph. The proven formula makes it possible to obtain mathematical models describing different processes on the graph. In this case, the resulting models are realized in the form of differential equations, defined pointwise, and supplemented by boundary conditions.

In [15] one will find interesting questions, in particular, on the theory of differential equations on graphs. These equations are used to describe reactions in chemical kinetics processes. Although this class of equations arises from chemical kinetics it is of independent mathematical interest.

6.2.2   The Concept of Quasi-fractal System Potential

Now let's introduce the concept of quasi-fractal system algebraic potential as a limit of the algebraic potential of a system of the level k of a quasi-fractal algebraic system at $k \to \infty$.

Definition 6.8 Let $G_S = A_1 = \langle A_1, \Omega_1 \rangle$ be a quasi-fractal group of factors that determine the system S.

Let's construct a flat graph of the lattice of subgroups at each level G_{S_k} with the number k of a quasi-fractal $G_S = A_1 = \langle A_1, \Omega_1 \rangle$. We arrange ordered pairs at the vertices of the graph (x_V, y_V), where x_V is a reciprocal of the number of synergistic effects of a subgroup of a group G_{Vk}, corresponding to vertex V, that is $pc(G_{Vk}) = \frac{1}{PC(G_{Vk})}$, and y_{Vk} is a communication level (maximum strength between G_V and subgroups of the group G_{S_k} incident to the vertex Vk in the lattice of subgroups of the group G_{S_k} which is defined as follows (Fig. 6.2):

We now assume in Definition 6.6 one by one $M = G_{Vk} \cup G_{Vk_i}$, where $i = 1, \ldots, n$, select the maximum value $pc(M) = \frac{1}{PC(M)}$.

$x = pc(G_{Vk})$, $y = pc(M)$, where $M = G_{Vk} \cup G_{Vk_i}$, $i = 1, \ldots, n$.

Fig. 6.2 A part of the lattice of subgroups of the group G_{S_k}, the level k of the quasi-fractal group of factors $G_S = A_1 = \langle A_1, \Omega_1 \rangle$

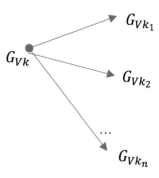

The difference $x - y$—analog of a distance between x и y. It can be done in accordance with the Definition 2.2, Chap. 2. After that, using the Lagrange theorem, we pull on the surface (curve) μk in the domain Ck to the points $\{ (x_{Vk}, y_{Vk}) | Vk$ runs over the set of vertices of the lattice graph of subgroups of the group $G_S \}$ and consider the integral

$$\int_{Ck} \frac{d\mu(y)}{|x - y|}$$

The limit of integrals $\lim\limits_{k \to \infty} \int_{Ck} \frac{d\mu(y)}{|x-y|}$ will be called the quasi-fractal system S potential according to the quasi-fractal model $G_S = A_1 = \langle A_1, \Omega_1 \rangle$.

Let's note, that if the system S is modeled by a fractal group of determining factors $G_S = A_1 = \langle A_1, \Omega_1 \rangle$, that is all of the groups G_{S_k} are isomorphic to each other for all levels k of a quasi-fractal system $G_S = A_1 = \langle A_1, \Omega_1 \rangle$ then fractal system $G_S = A_1 = \langle A_1, \Omega_1 \rangle$ potential coincides with the potential of the system S_k which is modeled by G_{S_k}.

6.2.3 Systems Classification by Quasi-fractal System Potential

In accordance with Definition 6.8 one gets the following systems classification by quasi-fractal system potential:

- systems of finite quasi-fractal potential,
- systems of infinite quasi-fractal potential,
- systems without quasi-fractal potential.

Definition 6.9 Let $G_S = A_1 = \langle A_1, \Omega_1 \rangle$ be a quasi-fractal group of factors that determine the system S. A system S is called a system of finite quasi-fractal potential if its quasi-fractal potential that is the limit of integrals (with accordance the Definition 6.8) $\lim\limits_{k \to \infty} \int_{Ck} \frac{d\mu(y)}{|x-y|}$ is finite.

Definition 6.10 Let $G_S = A_1 = \langle A_1, \Omega_1 \rangle$ be a quasi-fractal group of factors that determine the system S. A system S is called a system of infinite quasi-fractal potential if its quasi-fractal potential that is the limit of integrals (with accordance the Definition 6.8) $\lim\limits_{k \to \infty} \int_{Ck} \frac{d\mu(y)}{|x-y|}$ is infinite.

Definition 6.11 Let $G_S = A_1 = \langle A_1, \Omega_1 \rangle$ be a quasi-fractal group of factors that determine the system S. A system S is called a system of finite quasi-fractal potential if its quasi-fractal potential that is the limit of integrals (with accordance the Definition 6.8) $\lim\limits_{k \to \infty} \int_{Ck} \frac{d\mu(y)}{|x-y|}$ does not exist.

Examples

Let $G_S = A_1 = \langle A_1, \Omega_1 \rangle$ be a fractal group of factors that determine the system S. A fractal system S has quasi-fractal potential if there exists an integral

$$\int_C \frac{d\mu(y)}{|x-y|}$$

according to the Definition 6.7 because all levels of a fractal group $G_S = A_1 = \langle A_1, \Omega_1 \rangle$ are isomorphic to each other.

1. A fractal system S has finite quasi-fractal potential according to model $G_S = A_1 = \langle A_1, \Omega_1 \rangle$ if an integral

$$\int_C \frac{d\mu(y)}{|x-y|}$$

   is finite.
2. A fractal system S has infinite quasi-fractal potential according to model $G_S = A_1 = \langle A_1, \Omega_1 \rangle$ if an integral

$$\int_C \frac{d\mu(y)}{|x-y|}$$

   is infinite.
3. A fractal system S has no quasi-fractal potential if according to model $G_S = A_1 = \langle A_1, \Omega_1 \rangle$ if an integral

$$\int_C \frac{d\mu(y)}{|x-y|}$$

   does not exist.

One can make the following conclusion:
Potential can be considered as a measure of smart system reliability. See also [16].

6.3 Physical Meaning of Potential

In this section we try to validate that the physical meaning of the notion of potential is in the following: the potential of a system is a scale to measure the system capacity to change its properties while the system changes its structure.

In [17] one can find the following description of the notion of potential.

Potential generally refers to a currently unrealized ability, in a wide variety of fields from physics to the social sciences.

Potential Theory, in the initial meaning, was the doctrine of the properties of forces acting according the law of universal gravitation. In the formulation of this law, given by I. Newton (1687), we are talking only about the forces of mutual attraction acting on two material parts and small sizes, or material points. These forces are directly proportional to the product of masses of these particles and inversely proportional to the square of the distance between the particles. Therefore, the first and most important task from the point of view of celestial mechanics was to study the forces of attraction of a material point by a limited material body—an ellipsoid (because many celestial bodies have this particular shape).

Following [18] we mark that after the first achievements of Newton and other scientists, the works of J. Lagrange (1773), A. Legendre (1784–94) and P. Laplace (1782–99) were of primary importance here. Lagrange established that the field of gravitational forces is now said to be potential, and introduced a function that later J. P. Green (1828) called the potential function, and K. Gauss (1840)—just a potential.

Even K. Gauss and his contemporaries discovered that the method of potentials is applicable not only for solving problems of the theory of gravitation, but also for a wide range of problems of mathematical physics related, in particular, to electrostatics and magnetism. In Potential theory the main boundary-value problems were defined, such as the Dirichlet problem and the Neumann problem, the electrostatic problem of static charge distribution on conductors, and others. To solve these problems, in the case of domains with a sufficiently smooth boundary, varieties of potentials, i.e. special types of integrals that depend on parameters, such as the potential of volume-distributed masses, the potentials of simple and double layers, logarithmic potentials, etc., have proved to be an effective tool. The works of A. M. Lyapunov and V. A. Steklov of the late nineteenth century played a significant role in creating strict methods for solving the main boundary value problems. The study of the properties of potentials of various types has acquired a self-explanatory value in the theory.

The theory received a powerful incentive in the direction of generalization of the main tasks and completeness of the formulation of potential in the 1st half of twentieth century on the base of common notions of Measure theory and generalized functions. Modern Potential theory is closely related to the theory of analytical functions, harmonic functions, subharmonic functions, and probability theory. Modern period of development of Potential theory is characterized by the use of concepts and methods of topology and functional analysis, as well as abstract axiomatic methods.

Let's shortly enumerate main kinds of potential used in mathematics and physics:

Scalar potential, a scalar field whose gradient is a given vector field,

Vector potential, a vector field whose curl is a given vector field,
Potential function (disambiguation),
Potential variable (Boolean differential calculus),
Potential energy, the energy possessed by an object because of its position relative to other objects, stresses within itself, its electric charge, or other factors,
Magnetic potential, the vector potential or the magnetic scalar potential (ψ),
Electric potential, the amount of work needed to move a unit positive charge from a reference point to a specific point inside the field without producing any acceleration, Electromagnetic four-potential, a relativistic vector function from which the electromagnetic field can be derived,
Coulomb potential,
Van der Waals force, distance-dependent interactions between atoms or molecules,
Lennard-Jones potential, a mathematical model that approximates the interaction between a pair of neutral atoms or molecules,
Yukawa potential, a potential in particle physics which may arise from the exchange of a massive scalar field,
Gravitational potential.

6.3.1   Potential Theory in Social Sciences

In social sciences and in Knowledge Theory one can see the more advanced situation. The concept of the potential of the system is not only introduced into scientific circulation and theoretical calculations, but is used in practice in practical calculations. For example, in [19] it is marked that in recent years, there has been a trend towards more globalization of innovation and knowledge creation. It is also marked there that activities related to generation innovation and knowledge creation are more concentrated than other global activities. In terms of triadic patent application, United States, EU-28, Japan and China represented 86% of application. In terms of R&D spending and share of scientific publications, these economies represent 83% and 69%, respectively. These shares are notably larger than the share of the same economies in global trade, global GDP and global population [19].

These facts confirm the process of scientific potential notion usage the in practice.

Ibid [20] the technological potential in the development of the market was investigated, section Technology's potential for leapfrog solutions and market development in LDCs.

In Theory of Management such notion as fractal organization has appeared [21]. In [22], for example, a description of the fractal formation is given from a mathematical point of view, and also, a relationship between algebraic fractals and fractal organizations is carried out.

There are a lot of works devoted to fractal pedagogy [23, 24].

Let us return to the definition of the concept of "potential" in order to clarify its physical meaning. In [25], the following definition is given.

Definition 6.12 [25] Let μ be a be a positive Borel measure in R^3 such that $\int_{R^3} d\mu(y) < \infty$.

The potential of a measure μ is called a function

$$U^\mu(x) = \int\limits_{R^3} \frac{d\mu(y)}{|x - y|}$$

Let's remained the definition of Borel measure.

Definition 6.13 [26] Let X be a locally compact Hausdorff space, and let $\beta(X)$ be the smallest σ-algebra that contains the open sets of X; this is known as the σ—algebra of Borel sets. A Borel measure is any measure μ defined on the σ—algebra of Borel sets. Some authors require in addition that μ is locally compact, meaning that $\mu(C) < \infty$ for every compact set C. If a Borel measure μ is both inner regular and outer regular, it is called a regular Borel measure. If μ is both inner regular and locally finite, it is called a Radon measure.

In topology and related branches of mathematics, a Hausdorff space, separated space or T_2 space is a topological space where for any two distinct points there exists a neighborhood of each which is disjoint from the neighborhood of the other.

In [27] there is a comprehensive description of Borel sets.

6.3.2 Measure as a Scale

So, the physical meaning of the notion of potential is in the following: the potential of a system is a scale to measure the system capacity to change its properties while the system changes its structure.

Now we shall use the results of Chaps. 4 and 5.

Let's remind following Definitions 4.2 and 4.3 from Chap. 4.

For convenience let's again formulate Definitions 4.2 and 4.3 from Chap. 4

Definition 4.2, Chap. 4 Let $< A, \Omega >$ be a model and (B, Ω'), where $B \subseteq R$, be a numerical model. An ordered triple $< A, B, f >$, where $f = \langle f_1, f_2 \rangle$ is a homomorphfism from $< A, \Omega >$ into (B, Ω'), is called a scale.

Thus, measurement on a scale is the ordering (partial ordering) of a set in accordance with some property (measured on a given scale) and permissible scale transformations. From Definition 4.3 we have

Definition 4.3, Chap. 4 Let $< A, \Omega >$ be a model and (B, Ω'), where $B \subseteq R$, be a numerical model.

Let $< A, B, \mu >$ be a scale, where μ—is a Borel measure, which is a homomorphism, and (B, Ω'),

A homomorphism $\varphi : \langle R, \Omega', \leq \rangle \rightarrow \langle R, \Omega', \leq \rangle$, where $\leq$ is a relation "less or equal" on the set R of real numbers is called an admissible scale transformation of the scale $< A, B, \mu >$, if for any fixed homomorphism $f_0 :< A, \Omega > \rightarrow \langle B, \Omega' \rangle$ the following diagram

$$< A, \Omega > \xrightarrow{f_0} \langle B, \Omega' \rangle \xrightarrow{i} \langle R, \Omega', \leq \rangle$$

$$\mu \searrow \qquad\qquad \downarrow \varphi \qquad\qquad\qquad (6.1)$$

$$\langle B, \Omega' \rangle \xrightarrow{i} \langle R, \Omega', \leq \rangle$$

Is commutative one,
that is

$$\varphi i f_0 = i \mu :$$

$$< A, \Omega > \xrightarrow{f_0} \langle B, \Omega' \rangle \xrightarrow{i} \langle R, \Omega', \leq \rangle \qquad\qquad (6.2)$$

$$\mu \searrow \quad \varphi i f_0 = i \mu \quad \nearrow \varphi$$
$$\langle B, \Omega' \rangle \xrightarrow{i} \langle R, \Omega', \leq \rangle$$

where $i = \langle i_1, i_2 \rangle$ is a natural embedding $\langle B, \Omega' \rangle$ into $\langle R, \Omega', \leq \rangle$, that is $i_1 : B \rightarrow R$ is a natural embedding and $i_2 : \Omega' \rightarrow \Omega' \cup \{\leq\}$ is a natural embedding.

6.3.3 Probabilistic Measures that Define Measurement Scales

From the Chap. 5, Sect. 5.5.2 Homomorphisms that Define Measures, we get the following theorem.

Theorem 6.14 A finite Boolean algebra with a probability measure is a scale.
A finite σ-algebra with a probability measure is a scale (defines a scale).

6.4 Examples

Let's consider an example related to economics. Traditionally, periods of economic development are divided into the following ones:

The first period is characterized by the works of Walras containing descriptions of deterministic equilibria at a static level using systems of linear equations.

The second period was discovered by Pareto, in which the theory of comparative statics was developed.

The third period is determined by the works of Johnson, Slutsky, Hicks, Allen, in which the minimization of actions within the framework of an economic unit is characterized.

The fourth period is characterized by Samuelson's discovery of the correspondence principle, which states that the consequences of parametric changes in the system can be understood and predicted by comparing its state with the state of equilibrium.

At the fifth stage, the problem arises of creating a theory of comparative dynamics, that is, the behavior of the system and the changes occurring with it should be studied as a function of time, namely, the dependence of these changes on changes in time should be studied. From the point of view of mathematics, the occurrence of these periods can be described using the following scheme. Potential theory is a branch of mathematics and mathematical physics devoted to the study of the properties of partial differential equations in domains with a fairly smooth boundary by introducing special types of integrals depending on certain parameters called potentials [28].

In physics, field characteristics are inextricably linked with the concept of force. A force F acting on a given object will be proportional to the magnitude of the field vector at a given point in space. The relationship between the categories of field and power can be traced in socio-economic systems. Economic reality immanent to power, power relations between economic entities. From the point of view of property relations, the characteristics of competencies concentrated in the hands of each of the subjects of production relations are of particular importance. It is the structure and the size of this cluster that are critical for the formation of power relations between economic agents. In fact, this is nothing but the processes of concentration and centralization, that is, the enlargement of the scale of economic activity, the increase in production capacity and the relative size of firms within the industry or the unification of various enterprises under the control of one capital. An important source of changes in the current structure and links of financial and economic systems is the conflict of interests arising in connection with the separation of property from management and control in the modern economy [29].

In [30] there was built a dynamic model of duality of budget—tax system of Russia, using the analogue of the concept of physics that is the notion of system potential—the tax potential of the system. Let's recall that the potential energy Π of the system is part of the total energy of the system, depending on the relative position of the points that make up this system. The potential energy of the system is expressed in the form of a quadratic form and is determined by the initial and final state of the system. Kinetic energy of the system ΔT is determined by the speeds of the points of the system, that is, the difference in the states of the system in time. In classical mechanics, kinetic energy is the sum of the works:

$$\sum_{i=1}^{N} A_i^l + \sum_{i=1}^{N} A_i^k$$

where the first term $\sum_{i=1}^{N} A_i^l$ is determined by external forces acting on the system, and the second term $\sum_{i=1}^{N} A_i^k$ is determined by the internal forces of the system; the work is defined as the product of the force on the path. A conservative system is a system for which the sum of the kinetic energy T and the potential energy Π of the system is constant. The equilibrium state of a conservative system is described using the following system of second-order Lagrange differential equations,

$$\frac{d}{dt}\left(\frac{\partial T}{\partial \dot{q}_i}\right) - \frac{\partial T}{\partial q_i} = Q_i, i = 1, \ldots, N \tag{6.3}$$

or by equations in generalized Lagrangian coordinates [31–33],

$$\Delta T = \sum_{i=1}^{N} A_i^l + \sum_{i=1}^{N} A_i^k$$

where the dot denotes the time derivative, q_i are the generalized coordinates, the number of which is equal to the number N of the power-law freedom of the system, $\dot{q}_i$ are the generalized speeds, Q_i are the generalized forces, $i = 1, \ldots, N$. System (6.3) is a system of N ordinary second-order differential equations with unknown q_i—coordinates. The equations are invariant in form with respect to the choice of Lagrangian coordinates. Integrating these equations and determining the integration constants from the initial conditions, one finds $q_i(t); i - 1, \ldots, N$, that is, the law of motion of the system in generalized coordinates.

6.5 Conclusions

Potential is a measure of smart system reliability.

Potential of fractal smart system is a measure of fractal smart system reliability.

Quasi-fractal potential of a quasi-fractal smart system is a measure of a quasi-fractal smart system reliability.

References

1. Brelot, M.: Lectures on Potential Theory, (Reissued 1967), Notes by K.N. Gowrisankaran and M. K. Venkatesha Murthy, Second edition, revised and enlarged, with the help of S. Ramaswamy, Tata Institute of Fundamental Research, Bombay 1960, Reissued (1967)
2. Home page. https://ru.wikipedia.org/wiki/Potentialtheory
3. Home page. https://en.wikipedia.org/wiki/Energy
4. Home page. https://en.wikipedia.org/wiki/Measure
5. Targ, S.M.: Potential energy, Physical Encyclopedia, Ed. Prokhorov, A.M., Big Russian Encyclopedia, vol. 4. Moscow (1994)
6. Vechtomov, E.M.: Basic concepts of algebra. Publishing house Raduga-Press, Kirov (2013)

7. Serdyukova, N., Serdyukov, V.: Algebraic Formalization of Smart Systems. Theory and Practice, Smart Innovation, Systems and Technologies, SIST, vol. 91. Springer, Switzerland (2018)
8. Petrov AE, The Tensor Method of Dual Networks, International University of Nature, Society and Man, Dubna, Department of System Analysis and Management, Moscow (2007)
9. Home page. https://en.wikipedia.org/wiki/Gabriel_Kron
10. Kron, G.: A Short Course in Tensor Analysis for Electrical Engineers (1942), University Reprints (2019)
11. Kron, G.: Tensor analysis of networks, (1939), 2nd edn. (1965)
12. Kron, G.: Diakoptics—A piecewise solution of large-scale systems. Electr. J. 158–162, a serial of 20 chapters, from June 7, (1957) to February 13, London (1959)
13. Home page. https://en.wikipedia.org/wiki/Potential
14. Golovaneva, F.V., Lylov, E.V., Shabrov, S.A.: The Analog of the formula of integration by parts on a geometric graph, Bulletin of the Voronezh State University. Series: Phys. Maths 1, 82–86 (2018)
15. Volpert, A.I., Khudyaev, S.I.: Analysis in classes of discontinuous functions and equations of mathematical physics. Nauka, Moscow (1975). (in Russian)
16. Podgornaya, A.I., Khusainov, R.M.: Anti-crisis potential of fractal organizations. Bull. Kazan Technol. Univ. 16(20), 343–346 (2013)
17. Home page. https://en.wikipedia.org/wiki/Potential_(disambiguation)
18. Home page. https://bigenc.ru/Great Russian Encyclopedia
19. Financing for Development: Progress and Prospects, Report of the Inter-agency Task Force on Financing for Development, United Nations publication (2017)
20. LaFleur, M., Iversen, K., Lars Jensen, L.: Unlocking the potential of knowledge and technology for all, Department of Economic & Social Affairs, 1, APRIL United Nations publication (2018)
21. Warnecke H.-Y.: Revolution in der Unternehmerkultur. Fraktale Unternehmen/H. J. Warnecke, per. mit ihm. M.: Maik "Wissenschaft/interperiodika" (1999)
22. Shkarupeta, E.V., Smishlyaev, V.A.: Fractal Organizations in Conditions of Knowledge Economy, Home page. file:///C:/Users/пк/Downloads/fraktaln-e-organizatsii-v-usloviyah-ekonomiki-znaniy.pdf
23. Madzhuga, A., Sabekia, R., Sinitsyna, I., Salimova, R., Sadaeva, I.: Fractal pedagogy: theoretical and methodological preconditions of formation and development. Prof. Educ. Russia Abroad 2(22), 71–80 (2016)
24. Binsztok, A., Leja, K.: University as a fractal organization of knowledge, Wroclaw University of Economics, Gdansk University of Technology, In: Management Annual Conference on Higher Education Management and Development Annual Conference on Higher Education Management and Development in Central, Southern and Eastern Europe Danube University, Krems, 26–28 November (2006)
25. Wermer, J.: Potential Theory. Springer, Berlin, Heidelberg, New York (1974)
26. Home page. http://en.wikibedia.ru/wiki/Positive_Borel_measure#Formal_definition
27. Kanovei, V.G., Lyubetskiy, V.A.: The modern theory of sets: Borel and projective sets. MCCNMO, Moscow (2010). (in Russian)
28. Home page. https://en.wikipedia.org/wiki/Potential_theory
29. Verenikin, A.O.: Theory of economic potential and its analogue in mathematical physics, Economic problems of innovative development, Moscow State University, Moscow (2009)
30. Serdyukova, N.A.: Optimization of Tax System of Russia, Parts I and II. Budget and Treasury Academy, Rostov State Economic University, Moscow (2002) (in Russian)
31. Birkhoff, D.: Dynamical systems, editorial Board of the journal "Regular and chaotic dynamics", Publishing House "Udmurt University", Ischewsk (1999) (in Russian)
32. Gantmakher, F.R.: Theory of matrices, Home edition physical and mathematical literature, Moscow (1988)
33. Suslov, G.K.: Theoretical mechanics, Ogiz, State publishing house of technical and theoretical literature, Moscow, Leningrad (1946)

Chapter 7
System Structure Randomness and Integrity. Random Graphs Models

What is infrastructure? And this is connectedness!.
Raygorodsky. A., Models of Random Graphs,.
Moscow Publishing House, 2011, p. 43.

Abstract In this chapter the following basic questions are considered:

– random graphs,
– two Erdés–Renyi models,
– analogues of the Erdés–Renyi model for quasi-fractal graphs,
– quasi-fractal homomorphisms, scales, measures (continuation of Chap. 4),
– monitoring the functioning of a system represented by a quasi-fractal random graph,
– system structure randomness and integrity,
– infrastructure, giant component,
– the notion of random quasi-fractal graph (continuation of Chap. 4),
– random quasi-fractal scales (continuation of Chap. 4).

Keywords Random graphs models · Two Erdés–Renyi models · Infrastructure

7.1 Introduction

The main questions of this chapter will be the randomness of the structure of the system, the infrastructure of the system, its sustainability and the integrity of the system. So, we begin from the difference of the notions "structure" and "infrastructure". What does one understand as structure?

The word structure can refer to something that is made up of a number of components held or put together in a particular system. Structure also refer to the way in which parts are arranged together to form a complex whole. Structure is an internal organization of something, and infrastructure is a set of external elements that ensure the activity of something.

The word infrastructure has been used in French since 1875 and in English since 1887, originally meaning "The installations that form the basis for any operation

© Springer Nature Switzerland AG 2021
N. A. Serdyukova and V. I. Serdyukov, *Algebraic Identification of Smart Systems*, Intelligent Systems Reference Library 191, https://doi.org/10.1007/978-3-030-54470-6_7

or system" [5, 6]. The word was imported from French, where it was already used for establishing a roadbed of substrate material, required before railroad tracks or constructed pavement could be laid on top of it.

For understanding the difference between structure and infrastructure let's follow [1].

In accordance with [1], though the two terms structure and infrastructure are close related, there is a distinctive difference between structure and infrastructure. Structure is something that is made up of several parts put together in a particular system; it can also refer to the way in which these components are arranged together. Infrastructure refers to underlying base or foundation of a system or organization. This is the key difference between structure and infrastructure.

In accordance with [2], Infrastructure (lat. infra—"below", "under" and lat. structure—"structure", "location")—a complex of interconnected service structures or objects that make up and provide the basis for the functioning of the system [1]. In the modern language, in the common sense, infrastructure is a set of enterprises, institutions, management systems, communications, etc., providing the activities of the society or any of its spheres [2]. In English, the term "infrastructure" (Eng. infrastructure) appeared in 1920–1928 [2] and was originally used in the military sphere, where they designated a complex of structures that ensure the operation of the armed forces. Nowadays by infrastructure one understands the fundamental facilities and systems serving a country, city, or other area [1] including the services and facilities necessary for its economy to function [2], such as roads, railways, bridges, tunnels, water supply, sewers, electrical grids, and telecommunications (including Internet connectivity and broadband speeds). In general, it has also been defined as "the physical components of interrelated systems providing commodities and services essential to enable, sustain, or enhance societal living conditions" [2].

There are two general types of ways to view infrastructure, hard or soft. Hard infrastructure refers to the physical networks necessary for the functioning of a modern industry [2]. This includes roads, bridges, railways, etc. Soft infrastructure refers to all the institutions that maintain the economic, health, social, and cultural standards of a country [2]. This includes educational programs, official statistics, parks and recreational facilities, law enforcement agencies, and emergency services. Currently, the concept of infrastructure is expanding in terms of meaning. So recently the term infrastructure as code appeared. The infrastructure as code is the process of managing and provisioning computer data centers through machine-readable definition files, rather than physical hardware configuration or interactive configuration tools.

We are interested most of all in the mathematical sense of the words structure and infrastructure.

Definition 7.1 One can define a structure *StrS* of a system *S* as a system of links of the system *S* and an infrastructure *IStrS* of the system *S* as a system of links of the system *StrS*.

Such definition is correlated, for example, with the definition of the infrastructure as code that we have mentioned above.

According to [3, p. 43] by infrastructure the connectedness of the graph representing the structure of the system under study is understood. We have described in details systems links in [4, Chap. 5] using Algebraic Systems Theory and now we would like to concern some aspects of probabilities methods in studying this question.

7.2 Random Graphs

According to [5], in mathematics, random graph is the general term to refer to probability distributions over graphs. Random graphs may be described simply by a probability distribution, or by a random process which generates them [6]. The theory of random graphs lies at the intersection between graph theory and probability theory and uses methods of graph theory and of probability theory. From a mathematical perspective, random graphs are used to answer questions about the properties of typical graphs. Its practical applications are found in all areas in which complex networks need to be modeled. In a mathematical context, random graph refers almost exclusively to the Erdős–Rényi random graph model, which we shall use. Let's consider Erdős–Rényi definition [7].

7.2.1 History of the Erdés–Renyi Model of a Random Graph

Here we shall follow [8]. In accordance with [8], in the Graph Theory there exist two closely related models for description random graphs. These are Paul Erdos–Alfred Renyi model which was introduced first in 1959, and Edgar Gilbert model, also introduced in 1959. There exist two closely related Erdos –Renyi random graph models:

First Erdés–Renyi random graph model $G(n, M)$.
Second Erdés–Renyi random graph model $G(n, p)$.

In the first Erdés–Renyi random graph model $G(n, M)$, a graph is chosen uniformly at random from the collection of all graphs which have n nodes and M edges.

In the second Erdés–Renyi random graph model $G(n, M)$, each pair of n vertices $V = \{1, 2, \ldots, n\}$ are connected with the probability p independent from each other edge.

In the second Erdés–Renyi random graph model $G(n, p)$, a graph is constructed by connecting nodes randomly. Each edge is included in the graph with probability p independent from every other edge. Equivalently, all graphs with n nodes and M edges have equal probability of $p^M (1 - p)^{C_2^n - M}$.

The parameter p in this model can be thought of as a weighting function; as p increases from 0 to 1, the model becomes more and more likely to include graphs with more edges and less and less likely to include graphs with fewer edges. In particular, the case $p = 0.5$ corresponds to the case where all graphs on n vertices are chosen with equal probability.

The behavior of random graphs is often studied in the case where n, the number of vertices, tends to infinity. Although p and M can be fixed in this case, they can also be functions depending on n.

We shall use Erdés–Renyi second model.

Definition 7.2 [7] Let's fix an arbitrary natural number n and consider the set $V = \{1, \dots, n\}$—the set of all vertices of random graph. Only edges in the model will be random.

Let $C = C_n^2$ and $e_1, \dots, e_N$ are all possible edges that can be drawn on pairs of elements from V, that is $e_1, \dots, e_N$ are edges of the complete graph K_n. Let's set some $p \in [0, 1]$ and begin to select edges from the set $\{e_1, \dots, e_N\}$ according to the Bernoulli scheme with success probability p, that is in case of success, we include the next edge in the set of edges E under construction, and in case of failure, do not include it. As a result, we get a random graph $G = (v, E)$. As in the Bernoulli scheme, the resulting graph is the sequence $\omega = (x_1, \dots, x_N)$, consisting of zeros and ones, represented by the graph: $x_i = 1$ means $e_i \in E$; $x_i = 0$ means $e_i \notin E$. As a result one get the probability space $G(n, p) = \langle \Omega_n, F_n, P_n, p \rangle$ in which $|\Omega_n| = 2^N = 2^{C_n^2}$, $P_{n,p}(G) = p^{|E|} q^{C_n^2 - |E|}$. In other way, one can say that in the Erdés–Renyi model, each edge, regardless of all other edges, is included in a random graph with probability p. The probability space $G(n, p)$ is called the (second) Erdés–Renyi model. Each event from F_n is a set of graphs. One can also interpret events as graph properties. In fact: what does it mean to determine the probability that a graph has the property A? This means to describe the set A of those graphs that have this property and find the probability A. For example, if A is a set of all connected graphs then the probability of the event "a random graph is connected" is equal to $P_{n,p}(A) = \sum_{G \in A} P_{n,p}(G)$.

We now state these considerations more clearly.

7.3   Graphs' Probabilistic Properties

Let's continue to consider probabilistic constructions on graphs and algebraic systems. We shall follow [7].

7.3.1 Formal Description of the Probability Space Model

Let $V_n = \{v_1, \ldots, v_n\}, n = 1, 2 \ldots$ be a set of vertices. Let's construct a random graph on the set V_n. In addition, we would not consider graphs with multiple edges (multigraphs), graphs with loops (pseudographs), oriented graphs (digraphs). So, it turns out that the possible number of edges for a graph with n vertices is C_n^2. Let's connect any two vertices v_i and v_j with the probability $p_n \in [0; 1]$ independently of all other pairs of vertices. Thus, the edges appear in accordance with the Bernoulli independent tests' scheme tests with a probability of success p_n. So, we get a probability space $\langle \Omega_n, \Sigma_n, p_n \rangle$, in which
$\Omega_n = \{\Gamma_n = \langle V_n, E_n = \{(u_{v_i}, u_{v_j}) | i, j = 1, \ldots, n\}\}\}$, $\Sigma_n = 2^{\Omega_n}$ is the set of all subsets of the set Ω_n, $p_n(\Gamma_n) = p_n^{E_n}(1 - p_n)^{C_n^2 - |E_n|}$.

7.3.2 The "Almost All Graphs" Property

Let $G(n)$ be the set of all graphs with n vertices, P be some property given on the set of all graphs. We shall use the following designation:

$$G_P(n) = \{\Gamma \in G(n) | P(\Gamma)\}$$

Definition 7.3 [7] It is said that almost all graphs have the property P, if

$$\lim_{n \to \infty} \frac{|G_P(n)|}{|G(n)|} = 1,$$

and almost no graphs with property P, if

$$\lim_{n \to \infty} \frac{|G_P(n)|}{|G(n)|} = 0$$

The following theorems are well known.

Theorem 7.4 [9] Almost no Euler's graphs.
Almost all graphs are connected.
Almost all graphs are Hamilton's graph.
Now let's spread this definition to the case of P—pure properties of algebraic systems as follows. The notion of P—purities of algebraic systems was introduced in [10], was studied in [4]. It runs as follows.

Definition 7.5 A sub-algebra $\bar{B} = \langle B | \{f_\alpha^{n_\alpha} | \alpha \in \Gamma\}\rangle$ of an algebra $\bar{A} = \langle A | \{f_\alpha^{n_\alpha} | \alpha \in \Gamma\}\rangle$ is called P-pure in $\bar{A}$ or an embedding φ of a sub-algebra $\bar{B}$ into an algebra $\bar{A}$ is P-pure, if (1) every homomorphism $\bar{B} \xrightarrow{\alpha} \bar{C}$ of the subalgebra $\bar{B}$ into $\bar{C}$ (where $\bar{C}$ is an algebra of the signature $\{f_\alpha^{n_\alpha} | \alpha \in \Gamma\}$ of $\bar{A}$, and (2) $P(\bar{C})$ is true, (3)

P is a predicate on the class of algebras of the signature $\{f_\alpha^{n_\alpha} | \alpha \in \Gamma\}$ closed under taking subalgebras and factor algebras, can be continued to a homomorphism β of $\bar{A} = \langle A | \{f_\alpha^{n_\alpha} | \alpha \in \Gamma\}\rangle$ into $\bar{C} = \langle C | \{f_\alpha^{n_\alpha} | \alpha \in \Gamma\}\rangle$ in such a way that the following diagram is commutative:

$$0 \to \bar{B} = \langle B | \{f_\alpha^{n_\alpha} | \alpha \in \Gamma\}\rangle \overset{\varphi}{\to} \bar{A} = \langle A | \{f_\alpha^{n_\alpha} | \alpha \in \Gamma\}\rangle$$

$$\alpha \searrow \qquad \swarrow \beta \tag{7.1}$$

$$\bar{C} = \langle C | \{f_\alpha^{n_\alpha} | \alpha \in \Gamma\}\rangle$$

that is $\beta\varphi = \alpha$. (A note: The general operations of the same type in algebraic systems will be denoted in identical manner).

By parity of reasoning with the above definition, let's introduce the following notations.

Let P be any property given on the set of all algebraic systems $A = \langle A, \Omega\rangle$ of the signature Ω, $G(n, \Omega)$ be a set all algebraic systems of the signature Ω, main set of which consists of n elements, n is natural, $G_P(n, \Omega) = \{A \in G(n) | P(A)\}$.

Definition 7.6 It is said that almost all algebraic systems $A = \langle A, \Omega\rangle$ of the signature Ω have the property P, if

$$\lim_{n \to \infty} \frac{|G_P(n, \Omega)|}{|G(n, \Omega)|} = 1,$$

and almost no algebraic systems $A = \langle A, \Omega\rangle$ of the signature Ω with property P, if

$$\lim_{n \to \infty} \frac{|G_P(n)|}{|G(n)|} = 0$$

Since graphs as algebraic systems can be represented as semigroups the Definition 7.2 is in the frame of Definition 7.5.

Here the problem of finding an analog of the graph connectivity property for algebraic systems arises. We shall clear up a little the analogue of the graph connectivity property for algebraic systems in Chap. 8. Now we note that it concerns operations from the signature of the algebraic system, which become, with some probability, partially defined on the main set of the algebraic system.

7.3.3 Probabilistic Metrics. Connection with Elementary Theories

Let's now follow [11]. In graph theory, methods such as probability theory methods based on the Kolmogorov axiomatics, concepts and methods of the theory of algebraic systems, concepts and methods of measure theory, and closely related topologies and functional analysis have become widely used.

We need the concept of a probabilistic metric [11].

Let's use the following designations. Let $U = (U, d)$ be a complete separable metric space, $\langle \Omega, \Sigma, p \rangle$ be a probability space, $\mathcal{X} = \mathcal{X}(U)$ be a set of random variables given on probability space $\langle \Omega, \Sigma, p \rangle$, and taking values in $U = (U, d)$, $F_X(x)$ be a distribution function.

Firstly, let's remind the definition of Uniform metric, or Kolmogorov metric.

A uniform metric, or Kolmogorov metric, is defined as follows:

$$\rho(F_X, F_Y) = \sup\{|F_X(x) - F_Y(x)| | x \in R = R^1\}$$

Here x takes values in the one-dimensional Euclidean space $E = R^1$.

The question arises:

How one can set x on a quasi-fractal graph?—one can try as a characteristic function for a certain property, including a property expressed in a first-order language.

Now some words about the laws of zero and units for graphs, and the theorem from [7, p. 73], called the law of zero or one. We shall keep track the connection between this theorem and first-order language for graphs.

Theorem 7.7 [7, p. 73]

If $p = p(n)$ is a function such that $pn^\alpha \to \infty$ and $(1 - p)n^\alpha \to \infty$ for any $\alpha > 0$, and A is an arbitrary graph property, which can be expressed in a first-order language, then either almost surely property to A is satisfied, or almost certainly it does not take place. In other words, either $P_{n,p}(A) \to 1$, or $P_{n,p}(A) \to 0$.

It turns out that if p is a constant or function that does not tend to zero or unity too quickly, then any first-order property for a random graph has an asymptotic probability 0 or 1.

The following theorem gives an even more impressive result [7, p. 73; 12]

Theorem 7.8 [7, p. 73] (Theorem about connection with elementary theories)

If $p = n^{-\alpha}$, where $\alpha > 0$ is an irrational number, and A is an arbitrary property of the graph, which can be expressed in a first-order language, either almost certainly property A is satisfied, or almost certainly it does not take place. In other words, either $P_{n,p}(A) \to 1$, or $P_{n,p}(A) \to 0$.

Later on, we shall generalize these theorems to the case of a quasi-fractal graph (Sects. 7.4.1, Theorem 7.14, 7.15, Sect. 7.5, Theorem 7.18–7.21)

In Chap. 2 we have formulated the definition of a quasi-fractal graph. Let's repeat this definition now for convenience.

Definition 2.11 (Chapter 2, see also [13])

Let's consider graph $\Gamma_1 = \left\langle \{V_{\alpha_1}^1 | \alpha_1 \in \Lambda_1\}, \{u_{\alpha_1,\beta_1}^1 | \alpha_1, \beta_1 \in \Lambda_1\}\right\rangle$, where $\{V_{\alpha_1}^1 | \alpha_1 \in \Lambda_1\}$ is a set of vertices of a graph Γ_1, $\{u_{\alpha_1,\beta_1}^1 | \alpha_1, \beta_1 \in \Lambda_1\}$ is a set of edges of a graph Γ_1. Let's call a graph Γ_1 as a graph of the first level. Let, in turn $V_{\alpha_1}^1$ be a graph $\Gamma_2 = \left\langle \{V_{\alpha_2}^2 | \alpha_2 \in \Lambda_2\}, \{u_{\alpha_2,\beta_2}^2 | \alpha_2, \beta_2 \in \Lambda_2\}\right\rangle$, where $\{V_{\alpha_2}^2 | \alpha_2 \in \Lambda_2\}$ is a set of vertices of a graph Γ_2, $\{u_{\alpha_2,\beta_2}^2 | \alpha_2, \beta_2 \in \Lambda_2\}$ be a of edges of a graph Γ_2. Let's call a graph Γ_2 to be a graph of the second level. Let's continue this process by induction. If the graph $\Gamma_k = \left\langle \{V_{\alpha_k}^k | \alpha_k \in \Lambda_k\}, \{u_{\alpha_k,\beta_k}^k | \alpha_k, \beta_k \in \Lambda_k\}\right\rangle$, where $\{V_{\alpha_k}^k | \alpha_k \in \Lambda_k\}$ is a set of vertices of the graph Γ_k, $\{u_{\alpha_k,\beta_k}^k | \alpha_k, \beta_k \in \Lambda_k\}$ is the set of edges of the graph Γ_k, where Γ_k is a graph of the level k, then $V_{\alpha_k}^k, \alpha_k \in \Lambda_k$, is a graph $\Gamma_{k+1} = \left\langle \{V_{\alpha_{k+1}}^{k+1} | \alpha_{k+1} \in \Lambda_{k+1}\}, \{u_{\alpha_{k+1},\beta_{k+1}}^{k+1} | \alpha_{k+1}, \beta_{k+1} \in \Lambda_{k+1}\}\right\rangle$, where $\{V_{\alpha_{k+1}}^{k+1} | \alpha_{k+1} \in \Lambda_{k+1}\}$ is a set of vertices of the graph Γ_{k+1}, $\{u_{\alpha_{k+1},\beta_{k+1}}^{k+1} | \alpha_{k+1}, \beta_{k+1} \in \Lambda_{k+1}\}$ is a set of edges of the graph Γ_{k+1}. The graph Γ_{k+1} is called a graph of the level $k+1$. In this case we call the graph $\Gamma_1 = \left\langle \{V_{\alpha_1}^1 | \alpha_1 \in \Lambda_1\}, \{u_{\alpha_1,\beta_1}^1 | \alpha_1, \beta_1 \in \Lambda_1\}\right\rangle$ a quasi-fractal graph. If all the graphs $\Gamma_k = \left\langle \{V_{\alpha_k}^k | \alpha_k \in \Lambda_k\}, \{u_{\alpha_k,\beta_k}^k | \alpha_k, \beta_k \in \Lambda_k\}\right\rangle, r = 1, \ldots, n, \ldots$ are isomorphic to each other, then the graph $\Gamma_1 = \left\langle \{V_{\alpha_1}^1 | \alpha_1 \in \Lambda_1\}, \{u_{\alpha_1,\beta_1}^1 | \alpha_1, \beta_1 \in \Lambda_1\}\right\rangle$ is called a fractal graph. The graph $\Gamma_k = \left\langle \{V_{\alpha_k}^k | \alpha_k \in \Lambda_k\}, \{u_{\alpha_k,\beta_k}^k | \alpha_k, \beta_k \in \Lambda_k\}\right\rangle$, where $\{V_{\alpha_k}^k | \alpha_k \in \Lambda_k\}$ is a set of vertices of the graph Γ_k, $\{u_{\alpha_k,\beta_k}^k | \alpha_k, \beta_k \in \Lambda_k\}$ is a set of edges of the graph Γ_k, is called a quasi-fractal graph of the level k.

Thus, the notion of a fractal graph (respectively, quasi-fractal graph) determines, in fact, the structure of the simulated system, and, thus, the process of decomposition of the original system into subsystems. The concept of a quasi-fractal graph is a generalization of the notion of a graph, since a regular graph is a quasi-fractal graph of level 1.

We shall use the following notation for a quasi-fractal graph: $\Gamma_1 = \left\langle \{V_{\alpha_1}^1 | \alpha_1 \in \Lambda_1\}, \{u_{\alpha_1,\beta_1}^1 | \alpha_1, \beta_1 \in \Lambda_1\}, Q^1, W^1 \right\rangle = A_1 = \langle A_1; \Omega_1 \rangle$.

7.3.4 Sustainability Identification Level of a Quasi-fractal Graph

Now let's consider the graph as a dibasic algebraic system (that is, as an algebraic system with two main sets—the set of vertices and the set of edges, and the two-place predicates Q and W, given respectively on the set of vertices of the graph and on the set of edges of the graph.

Let's consider that $Q(a_i, a_j) = 1$, if there exists an edge of a graph which connects vertices a_i, a_j, and $Q(a_i, a_j) = 0$, if not. Let's assume that the predicate W is true on edges incident to the same vertex, and false otherwise. Next, we generalize this construction to the case of a quasi-fractal graph, that is, we shall consider a quasi-fractal graph as a quasi-fractal algebraic system, at each level of which there is a dibasic algebraic system with predicates. Note that in this case, the set of predicates Q and W can be extended and supplemented by other predicates at each level of the quasi-fractal.

Further we apply the fixed point as in Chap. 3, to the quasi-fractal algebraic system $\Gamma_1 = \left\langle \{V_{\alpha_1}^1 | \alpha_1 \in \Lambda_1\}, \{u_{\alpha_1, \beta_1}^1 | \alpha_1, \beta_1 \in \Lambda_1\}, Q^1, W^1 \right\rangle = A_1 = \langle A_1; \Omega_1 \rangle$.

Let's define the metrics d on the quasi-fractal $\Gamma_1 = \left\langle \{V_{\alpha_1}^1 | \alpha_1 \in \Lambda_1\}, \{u_{\alpha_1, \beta_1}^1 | \alpha_1, \beta_1 \in \Lambda_1\}, Q^1, W^1 \right\rangle = A_1 = \langle A_1; \Omega_1 \rangle$ in accordance to Examples of Chap. 2. According to [14], this complete metric compactum, in the case when its dimension is n, s homeomorphic to the unit ball D_n in R^n, where $D_n = \{x \in R^n | \|x\| \leq 1\}$ and $|x| = |x_1, \ldots, x_n| = \left(x_1^2, \ldots, x_n^2\right)^{\frac{1}{2}}$. Then, according to Brouwer fixed—point theorem, we obtain a fixed point of the compressing map from Examples of Chap. 2, which defines the level of sustainability identification of a quasi-fractal graph $\Gamma_1 = \left\langle \{V_{\alpha_1}^1 | \alpha_1 \in \Lambda_1\}, \{u_{\alpha_1, \beta_1}^1 | \alpha_1, \beta_1 \in \Lambda_1\}, Q^1, W^1 \right\rangle = A_1 = \langle A_1; \Omega_1 \rangle$.

7.4 Monitoring the Functioning of a System Represented by a Quasi-fractal Random Graph

We shall use Erdös–Renyi models and algorithms to study the sustainability of the system S, modeled by a quasi-fractal graph.

Definition 7.9 Let system S is simulated by a quasi-fractal algebraic system $A_1 = \langle A_1; \Omega_1 \rangle$. We define a random graph on each level k of this quasi-fractal system as follows:
$\Gamma_k = \left\langle \{V_{\alpha_k}^k | \alpha_k \in \Lambda_k\}, \{u_{\alpha_k, \beta_k}^k | \alpha_k, \beta_k \in \Lambda_k\} \right\rangle$, where $\{V_{\alpha_k}^k | \alpha_k \in \Lambda_k\}$ is a set of vertices of the graph Γ_k, $\{u_{\alpha_k, \beta_k}^k | \alpha_k, \beta_k \in \Lambda_k\}$ is the set of edges of the graph Γ_k. Here an element $V_{\alpha_k}^k$ is a vertice of a graph Γ_k, corresponding to an element $\alpha_k \in \Lambda_k$ of an algebraic system $A_k^\alpha = \langle A_k^\alpha; \Omega_k \rangle, \alpha \in \Lambda_k$, where $A_k^\alpha = \langle A_k^\alpha; \Omega_k \rangle, \alpha \in \Lambda_k$, is an algebraic system of the level k of the quasi-fractal algebraic system. Let's assume the elements $\alpha_k, \beta_k \in A_k^\alpha$ interact in the system S with the probability p_k. Then the elements $V_{\alpha_k}^k$ and $V_{\beta_k}^k$ are connected by the edge u_{α_k, β_k}^k with the probability p_k. Let's call a quasi-fractal graph $\Gamma_1 = \left\langle \{V_{\alpha_1}^1 | \alpha_1 \in \Lambda_1\}, \{u_{\alpha_1, \beta_1}^1 | \alpha_1, \beta_1 \in \Lambda_1\} \right\rangle = A_{rqf} = \langle A_{rqf}; \Omega_{rqf} \rangle$ which corresponds to this construction a random quasi-fractal graph of the first sort, determining the functioning of the system determining the

functioning of the system S. Here the designation $\boldsymbol{A_{rqf}} = \langle A_{rqf}; \Omega_{rqf} \rangle$ plays the role of a universal variable to denote a random quasi-fractal algebraic system.

A quasi-fractal algebraic system $\boldsymbol{A_1} = \langle A_1; \Omega_1 \rangle$ is called a finite quasi-fractal system if, at each level of the quasi-fractal k the algebraic system $A_k^\alpha = \langle A_k^\alpha; \Omega_k \rangle$, $\alpha \in \Lambda_k$, is finite.

Now we shall follow the following two theorems of Erdös–Renyi about system functioning (about system integrity).

Theorem 7.10 [7, Theorem 13, p. 45]

Let $p = \frac{clnn}{n}$. If $c \geq 3$, then almost certainly a random graph is connected. If $c \geq 1$, then then almost certainly a random graph is not connected.

Theorem 7.11 [7, Theorem 14, p. 45]

Let $p = \frac{clnn}{n}$. If $c \geq 3, n \geq 100$, then $P_{n,p}(G \text{ is connected}) > 1 - \frac{1}{n}$.

7.4.1 Description of the Erdés–Renyi Model of a Random Quasi-fractal Finite Graph

In accordance with [7] for every level k let's consider the set $V\{_{\alpha_k}^k|\alpha_k \in \Lambda_k\} = \{1, \ldots, n_k\}$ of vertices of a random graph. Only edges will be random in the graph. Let $N_k = C_{n_k}^2$ and $\{e_1, \ldots, e_{n_k}\}$ be edges of a complete graph K_{n_k}. Let $p_k \in [0, 1]$. Let's begin to choose edges from the set $V\{_{\alpha_k}^k|\alpha_k \in \Lambda_k\} = \{1, \ldots, n_k\}$ according to Bernoulli scheme with probability of success p_k, that is, if successful, we include the next edge in the set of edges under construction $E_k = \left\{ u_{\alpha_k,\beta_k}^k |\alpha_k, \beta_k \in \Lambda_k \right\}$ of the random graph $\Gamma_k = \left\langle \{V_{\alpha_k}^k|\alpha_k \in \Lambda_k\}, \left\{ u_{\alpha_k,\beta_k}^k |\alpha_k, \beta_k \in \Lambda_k \right\} \right\rangle$, and in case of failure—do not include.

It turns out that, as in the Bernoulli scheme, a graph is a sequence $\omega = (x_1, \ldots, x_{N_k})$ of zeros and units. In fact, the sequence is represented by the graph: $x_i = 1$ means $e_i \in E_k$, if $x_i \neq 1$ then $e_i \notin E_k$. So, we get the probability space $G(n_k, p_k) = (\Omega_{n_k}, \mathcal{F}_{n_k}, \mathcal{P}_{n_k,p_k})$, in which

$$|\Omega_{n_k}| = 2^{N_k} = 2^{C_{n_k}^2}, \quad \mathcal{P}_{n_k,p_k} = p_k^{|E_k|}(1 - p_k)^{C_{n_k}^2 - |E_k|}$$

Thus, in the Erdös–Renyi model, each edge, independently of all other edges, is included in a random graph of level k with probability p_k. Each event from $\mathcal{F}_{n_k}$ is a set of graphs. So, it is natural to interpret events as properties of graphs. For example: what does it mean to calculate the probability that a graph has a given property? This means that it is required to describe the set A of those graphs that possess the given property, and then find the probability A. If A is a set of all connected graphs, then the probability of the event "the random graph is connected" is equal to

$$\mathcal{P}_{n_k,p_k}(A) = \sum_{G \in A} \mathcal{P}_{n_k,p_k}(G)$$

In [7, p. 42] the following theorem is proved:

Theorem 7.12 [7, p. 42]
Let $p = \frac{clnn}{n}$. If $c > 1$, then almost certainly a random graph is connected. If $c < 1$, then almost certainly a random graph is not connected.

Theorem 7.13 [7] (network reliability theorem). Let $p = \frac{clnn}{n}$. If $c \geq 3, n \geq 100$, then

$$\mathcal{P}_{n,p}(G \text{ is connected}) > 1 - \frac{1}{n}$$

We obtain the following theorems describing the properties of quasi-fractal random graphs here from.

Theorem 7.14 Let the system S is simulated by a finite quasi-fractal algebraic system $A_1 = \langle A_1; \Omega_1 \rangle$ and by a random quasi-fractal finite graph of the first sort $\Gamma_1 = \left\langle \{V_{\alpha_1}^1 | \alpha_1 \in \Lambda_1\}, \{u_{\alpha_1,\beta_1}^1 | \alpha_1, \beta_1 \in \Lambda_1\} \right\rangle$, which defines the system S process functioning. Let for any natural k the following equality holds:

$$p_k = \frac{clnk}{k}$$

If $c > 1$, then almost certainly a random graph.
$\Gamma_k = \left\langle \{V_{\alpha_k}^k | \alpha_k \in \Lambda_k\}, \{u_{\alpha_k,\beta_k}^k | \alpha_k, \beta_k \in \Lambda_k\} \right\rangle$ is connected, that is a system S retains integrity property. If $c < 1$, then almost certainly a random graph $\Gamma_k = \left\langle \{V_{\alpha_k}^k | \alpha_k \in \Lambda_k\}, \{u_{\alpha_k,\beta_k}^k | \alpha_k, \beta_k \in \Lambda_k\} \right\rangle$ is not connected, that is a system S loses integrity property.

Theorem 7.15 (theorem on the reliability of a system modeled by a finite quasi-fractal algebraic system) Let a system S is simulated by a quasi-fractal finite algebraic system $A_1 = \langle A_1; \Omega_1 \rangle$ and by a quasi-fractal finite random graph $\Gamma_1 = \left\langle \{V_{\alpha_1}^1 | \alpha_1 \in \Lambda_1\}, \{u_{\alpha_1,\beta_1}^1 | \alpha_1, \beta_1 \in \Lambda_1\} \right\rangle$ of the first sort defining the system S process functioning. Let for every natural fixed k the following equality holds:

$$p_k = \frac{clnk}{k}$$

If $c \geq 3, k \geq 100$, then

$$\mathcal{P}_{n_k,p_k}\left(\Gamma_k = \langle \{V_{\alpha_k}^k | \alpha_k \in \Lambda_k\}, \{u_{\alpha_k,\beta_k}^k | \alpha_k, \beta_k \in \Lambda_k\} \rangle \text{ is connected}\right) > 1 - \frac{1}{k}.$$

7.5   System Structure Randomness and Integrity. Infrastructure and Connectiveness. Quasi-fractal Connectiveness

Definition 7.16 Let system S is simulated by a quasi-fractal finite algebraic system $A_1 = \langle A_1; \Omega_1 \rangle$. Let every level of a quasi-fractal finite graph corresponding to a system $A_1 = \langle A_1; \Omega_1 \rangle$ exists with a probability p. That is at each level k of this quasi-fractal system the graph $\Gamma_k = \left\langle \{ V_{\alpha_k}^k | \alpha_k \in \Lambda_k \}, \{ u_{\alpha_k, \beta_k}^k | \alpha_k, \beta_k \in \Lambda_k \} \right\rangle$ with the set of vertices $\{ V_{\alpha_k}^k | \alpha_k \in \Lambda_k \}$ and with the set of edges $\{ u_{\alpha_k, \beta_k}^k | \alpha_k, \beta_k \in \Lambda_k \}$ is defined. Here $V_{\alpha_k}^k$ is the vertice of the graph Γ_k, which corresponds to the element $\alpha_k \in \Lambda_k$ of the algebraic system $A_k^\alpha = \langle A_k^\alpha; \Omega_k \rangle, \alpha \in \Lambda_k$, of a fractal level k. We assume that the elements $\alpha_k, \beta_k \in A_k^\alpha$ interact in the system S. Then the elements $V_{\alpha_k}^k$ and $V_{\beta_k}^k$ are connected by the edge u_{α_k, β_k}^k. The quasi-fractal graph $\Gamma_1 = \left\langle \{ V_{\alpha_1}^1 | \alpha_1 \in \Lambda_1 \}, \{ u_{\alpha_1, \beta_1}^1 | \alpha_1, \beta_1 \in \Lambda_1 \} \right\rangle$ which corresponds to this construction is called a quasi-fractal graph of the second sort defining system S functioning.

Definition 7.17 Let system S is simulated by a quasi-fractal finite algebraic system $A_1 = \langle A_1; \Omega_1 \rangle$. Let every level of the quasi-fractal graph defining system $A_1 = \langle A_1; \Omega_1 \rangle$ functioning exists with the probability p. At each level k of this quasi-fractal system the graph $\Gamma_k = \left\langle \{ V_{\alpha_k}^k | \alpha_k \in \Lambda_k \}, \{ u_{\alpha_k, \beta_k}^k | \alpha_k, \beta_k \in \Lambda_k \} \right\rangle$ with the set of vertices $\{ V_{\alpha_k}^k | \alpha_k \in \Lambda_k \}$ and a set of edges $\{ u_{\alpha_k, \beta_k}^k | \alpha_k, \beta_k \in \Lambda_k \}$ is defined. Here an element $V_{\alpha_k}^k$ is a vertice of the graph Γ_k, which corresponds to an element $\alpha_k \in \Lambda_k$ of the algebraic system $A_k^\alpha = \langle A_k^\alpha; \Omega_k \rangle, \alpha \in \Lambda_k$, of a fractal level k. Let's assume that the elements $\alpha_k, \beta_k \in A_k^\alpha$ interact in the system S with the probability p_k. Then elements $V_{\alpha_k}^k$ and $V_{\beta_k}^k$ are connected by an edge u_{α_k, β_k}^k with the probability p_k. The quasi-fractal graph $\Gamma_1 = \left\langle \{ V_{\alpha_1}^1 | \alpha_1 \in \Lambda_1 \}, \{ u_{\alpha_1, \beta_1}^1 | \alpha_1, \beta_1 \in \Lambda_1 \} \right\rangle$ which corresponds to this construction is called a quasi-fractal graph of the third sort defining system S functioning.

From the Erdös–Renyi model we obtain the following theorems.

Theorem 7.18 Let the system S is simulated by a quasi-fractal finite system $A_1 = \langle A_1; \Omega_1 \rangle$ and by a random quasi-fractal finite graph $\Gamma_1 = \left\langle \{ V_{\alpha_1}^1 | \alpha_1 \in \Lambda_1 \}, \{ u_{\alpha_1, \beta_1}^1 | \alpha_1, \beta_1 \in \Lambda_1 \} \right\rangle$, of the second sort defining system S functioning. Let for every natural fixed k

$$p = \frac{c \ln k}{k}$$

If $c > 1$, then almost certainly a random graph $\Gamma_k = \left\langle \{ V_{\alpha_k}^k | \alpha_k \in \Lambda_k \}, \{ u_{\alpha_k, \beta_k}^k | \alpha_k, \beta_k \in \Lambda_k \} \right\rangle$ is connected that is the system S retains integrity property. If $c < 1$, then almost certainly a random graph

$\Gamma_k = \left\langle \left\{ V_{\alpha_k}^k | \alpha_k \in \Lambda_k \right\}, \left\{ u_{\alpha_k, \beta_k}^k | \alpha_k, \beta_k \in \Lambda_k \right\} \right\rangle$ is not connected, that is the system S loses fractal property (fractal integrity).

Theorem 7.19 (theorem on the reliability of a quasi-fractal structure) Let the system S is simulated by a quasi-fractal finite system $A_1 = \langle A_1; \Omega_1 \rangle$ and by a random quasi-fractal finite graph $\Gamma_1 = \left\langle \left\{ V_{\alpha_1}^1 | \alpha_1 \in \Lambda_1 \right\}, \left\{ u_{\alpha_1, \beta_1}^1 | \alpha_1, \beta_1 \in \Lambda_1 \right\} \right\rangle$, of the second sort defining system S functioning. Let for every natural fixed k

$$p = \frac{clnk}{k}$$

If $c \geq 3, n \geq 100$, then

$$\mathcal{P}_{n_k, p_k}\left(\Gamma_k = \left\langle \left\{ V_{\alpha_k}^k | \alpha_k \in \Lambda_k \right\}, \left\{ u_{\alpha_k, \beta_k}^k | \alpha_k, \beta_k \in \Lambda_k \right\} \right\rangle \text{ is connected} \right) > 1 - \frac{1}{n}.$$

Theorem 7.20 Let the system S is simulated by a quasi-fractal finite system $A_1 = \langle A_1; \Omega_1 \rangle$ and by a random quasi-fractal finite graph $\Gamma_1 = \left\langle \left\{ V_{\alpha_1}^1 | \alpha_1 \in \Lambda_1 \right\}, \left\{ u_{\alpha_1, \beta_1}^1 | \alpha_1, \beta_1 \in \Lambda_1 \right\} \right\rangle$, of the third sort defining system S functioning. Let for every natural fixed k

$$p_k = \frac{clnk}{k}, \ p = \frac{c'lnk}{k}$$

If $c > 1, c' > 1$, then almost certainly a random graph
$\Gamma_k = \left\langle \left\{ V_{\alpha_k}^k | \alpha_k \in \Lambda_k \right\}, \left\{ u_{\alpha_k, \beta_k}^k | \alpha_k, \beta_k \in \Lambda_k \right\} \right\rangle$ is connected that is the system S retains integrity property. If $c < 1$, or $c' < 1$ then almost certainly a random graph.
$\Gamma_k = \left\langle \left\{ V_{\alpha_k}^k | \alpha_k \in \Lambda_k \right\}, \left\{ u_{\alpha_k, \beta_k}^k | \alpha_k, \beta_k \in \Lambda_k \right\} \right\rangle$ is not connected, that is the system S loses integrity property.

Theorem 7.21 Let the system S is simulated by a quasi-fractal finite system $A_1 = \langle A_1; \Omega_1 \rangle$ and by a random quasi-fractal finite graph $\Gamma_1 = \left\langle \left\{ V_{\alpha_1}^1 | \alpha_1 \in \Lambda_1 \right\}, \left\{ u_{\alpha_1, \beta_1}^1 | \alpha_1, \beta_1 \in \Lambda_1 \right\} \right\rangle$, of the third sort defining system S functioning. Let for every natural fixed k

$$p_k = \frac{clnk}{k}, \ p = \frac{c'lnk}{k}$$

$$p_k = \frac{clnk}{k}, \ p = \frac{c'lnk}{k}$$

If $c \geq 3, n \geq 100$, then

$$\mathcal{P}_{n_k, p_k}\left(\Gamma_k = \left\langle \left\{ V_{\alpha_k}^k | \alpha_k \in \Lambda_k \right\}, \left\{ u_{\alpha_k, \beta_k}^k | \alpha_k, \beta_k \in \Lambda_k \right\} \right\rangle \text{ is connected} \right) > 1 - \frac{1}{n}.$$

If $c' \geq 3$, $n \geq 100$, then

$$\mathcal{P}_{n_k, p_k}\left(\Gamma_k = \left\langle\left\{V_{\alpha_k}^k | \alpha_k \in \Lambda_k\right\}, \left\{u_{\alpha_k, \beta_k}^k | \alpha_k, \beta_k \in \Lambda_k\right\}\right\rangle \text{ is connected}\right) > 1 - \frac{1}{n}.$$

7.5.1 Infrastructure. Giant Component

Let's now concern the notion of a giant component. The next theorem was obtained from the second Erdös–Renyi random graph model $G(n, p)$ and contains even deeper information about the nature of reliability and connectivity [15–17].

Theorem 7.22 [15–17]

Let's consider the second Erdös–Renyi random graph model $G(n, p)$. Let $p = \frac{c}{n}$. If $c < 1$ then there exists such a constant $\beta = \beta(c)$ that almost always the size of each connected component of a random graph does not exceed βlnn. If $c > 1$ then there exists such a constant $\gamma = \gamma(c)$ that almost always in a random graph there is exactly one component of size $\geq \gamma n$, where

$$G(n, p) = \left\langle \Omega_n, F_n, P_{n,p}\right\rangle, \Omega_n = \{G = (V_n, E), F_n = 2^{\Omega_n}, P_{n,p}(G) = p^{|E|}(1 - p)^{C_n^2 - |}$$

This component is called a giant component.

Examples

The examples will concern the notion of infrastructure [7]

Let's concern again the notion of infrastructure and consider the following example connected with the notion of infrastructure from [7]: What is infrastructure? And this is connectedness!

Example 1. Destroying net connections We "allow" the enemy to destroy every path with a probability of 0.99 and, nevertheless, maintain the integrity of the road network with a probability of 0.999! "If we return to the "strategic" interpretation of the connectedness problem, we can say so. "Allowing" the enemy to destroy each railway track with a probability of $1 - \frac{1}{n}$, with a probability close to unity, we will save, if not the whole country, then at least its huge piece—a piece of size γn. This piece is called the giant component. If we give the enemy even more freedom, then the collapse is almost inevitable, and this is a catastrophic collapse: the islands of connectivity most likely have a negligible—logarithmic—size. By the way, even in the presence of a giant component, all other components are tiny—logarithmic. This can also be rigorously substantiated. In principle, this is the structure of any large network. For example, this is how the Internet works. In it, vertices are sites, and links are links between sites. There is a giant cohesive component and a mass of small, scattered "margins".

Example 2. Giant component in states' development in time scale One more strategic interpretation from [7] which concern state's development Let's imagine that p is time changes in the interval $[0, 1]$. If a sufficiently large n is given, then for $p \ll 1$ we have feudal fragmentation, time passes and for $p \geq \frac{1}{n}$ an empire arises; finally, with $p \gg \frac{lnn}{n}$ the empire gains world domination.

Example 3. Giant component in knowledge system Let's now consider the following example associated with the knowledge system. The integrity of the student's knowledge system maintaining. We can interpret a knowledge system S as a net [18]

Let's begin from the notion of formalization of the concept of "task" in the training system

Example 4. Formalization of the concept of "task" in the training system Let the knowledge system S be represented as the graph. A task is a path in this graph. Recall the definition of a path in a graph. A path in a graph is a sequence of vertices in which each vertex is connected to the next one by an edge.

Definition 7.23 [19, 20]
Let Γ be an undirected graph. A path in Γ is such a finite or infinite sequence of edges and vertices $w = (\ldots, a_0, E_0, a_1, E_1, \ldots, E_{n-1}, a_n, \ldots)$, that every two adjacent edges E_{i-1}, E_i have a common vertex a_i.

The set of tasks in the knowledge system S will be denoted as $T = \{T_i | i \in I\}$, the path corresponding to the task $T_i, i \in I$, will be denoted as $W(T_i), i \in I$, and a set of paths in the graph Γ, corresponding to the set of tasks $T = \{T_i | i \in I\}$, will be denoted as $W(T) = \{W(T_i) | i \in I\}$.

One can formalize the knowledge, skills and abilities of a particular student using the notion of graph homomorphism.

Formalization of the concepts of "knowledge", "skills", "skills" of a particular student in the knowledge system S

Let $\{\alpha | \alpha \in \Lambda\}$ be a set of students of the knowledge system S.

Definition 7.24 Let $f_\alpha : \Gamma \to \Gamma$ be a homomorphism of a graph Γ into itself. Then the image $f_\alpha(\Gamma)$ is a the system of knowledge, abilities, skills of the student $\alpha, \alpha \in \Lambda$, the image $f_\alpha(\{W(T_i) | i \in I\})$ is the ability to solve the tasks of the student $\alpha, \alpha \in \Lambda$.

Recall that a homomorphism of a graph Γ is a mapping $f : \Gamma \to \Gamma'$ of vertices of a graph Γ into a set of vertices of a graph Γ' under which the incidence relation is preserved. Let's show now how to formalize attempts to solve the problem by students. To do this, we embed the graph Γ representing the knowledge system S into the complete graph Γ_S. Then we construct the free group G_Γ such that its Van Kampen diagram corresponds to the graph Γ_S.

Let's use the way $W(T_i), i \in I$, in the graph Γ_S and set off corresponding to $W(T_i), i \in I$, part $D(G_\Gamma)$ n the Van Kampen diagram and $H(G_\Gamma)$ in the group G_Γ, then close $H(G_\Gamma)$ up to the subgroup $G(W(T_i))$, which we call the set of attempts to solve the problem T_i. After that we build a homomorphism $f_i : G_\Gamma \to G(W(T_i))$.

Then the kernel of this homomorphism $Ker f_i$ is a measure of the deviation of the image of the student's knowledge system from the knowledge system. The random system $f_\alpha(\{W(T_i)|i \in I\})$ should be a connected graph for a student α to be successful.

So, we are in the conditions of Theorem 7.22.

The question arises: how to go to infinite random graphs?

7.5.2 Infinite Random Graphs

Following [21], we continue to study Erdős–Rényi model of the second sort.

If we start with an infinite number of vertices and choose each possible edge independently with probability $0 < p < 1$, we get an object Γ called an infinite random graph. Except for trivial cases when p is equal 0 or 1, such graph Γ almost reliably possesses the following properties: If any $n + m$ elements are given $a_1, \ldots, a_n, b_1, \ldots, b_m$ there exists a vertex c of the graph Γ, adjacent to each vertex $a_1, \ldots, a_n$, and not connected to any vertex $b_1, \ldots, b_m$.

It turns out that if the set of vertices of the graph Γ is countable, then, up to isomorphism, there exists a unique graph with such properties, namely, the Rado graph. The Rado graph is the unique (up to isomorphism) countable graph R such that for any finite graph Γ and its vertex any embedding Γ—v in R as a generated subgraph can be extended to an embedding Γ in R.

So, every finite or countably infinite graph is an induced subgraph of the Rado graph, and can be found as an induced subgraph by a greedy[1] algorithm that builds up the subgraph one vertex at a time. The Rado graph is uniquely defined, among countable graphs, by an extension property that guarantees the correctness of this algorithm: no matter which vertices have already been chosen to form part of the induced subgraph, and no matter what pattern of adjacencies is needed to extend the subgraph by one more vertex, there will always exist another vertex with that pattern of adjacencies that the greedy algorithm can choose.

As a result, the Rado graph (Fig. 7.1) contains all finite and countable infinite graphs as subgraphs.

Thus, any countable infinite graph is almost certainly a Rado graph, which for this reason is sometimes called simply a random graph. A similar result is not true for uncountable graphs for which there are many (non-isomorphic) graphs satisfying the above condition.

Another model generalizing the Hilbert model of a random graph is a graph model of a random scalar product. A random scalar product graph associates a real vector with each vertex. The probability of the edge (u, v) between any vertices u and v is some function of the scalar product $u \cdot v$ of the vectors corresponding to them.

---

[1]Is any algorithm that follows the problem-solving heuristic of making the locally optimal choice at each stage with the intent of finding a global optimum. This heuristic does not intend to find a best solution, but it terminates in a reasonable number of steps; finding an optimal solution to such a complex problem typically requires unreasonably many steps.

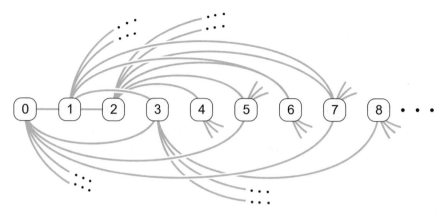

Fig. 7.1 Rado graph [21]

7.5.3 Target Subsystem of the System. The Hypothesis

The following hypothesis arises: the giant component is the target subsystem of the system.

The system's description using a model of factors which determine the system [4], allows one to make the following conclusions:

1. We identify the system by its effect on the external environment, that is by its functioning or by the functioning of the factors determining the system. So, one of the system's goals is its functioning. The system's functioning is ensured by its system of connections, that is by system's infrastructure. From here we come that the giant component of the system (that is the part of the system that preserves connectivity or infrastructure according to the second Erdős–Rényi model) is the target subsystem of the system.
2. The system's description with the usage of the model of factors which determine the system, shows one that, from the algebraic formalization of systems' point of view, any system other than non-functioning systems (that is, for example, systems in equilibrium state, (for example, a system described by a unit group) is complicated one.
3. The system as a result of its functioning, comes to one of its possible final states $\{S_1, S_2, \ldots, S_n\}$ in accordance with the model of factors determining the system selected for its description. The giant components of the subsystems $S_1, S_2, \ldots, S_n$ will be called the target subsystems of the system S.

Let's note that Gödel's incompleteness theorem justifies the possible use of different models of factors determining the system.

7.6 Quasi-fractal Random Scales

The following question arises: to what degree will the scale, which has a quasi-fractal construction, preserve its measuring properties during destruction?

7.6.1 Random Quasi-fractal Graph

For convenience, let's give Definition 2.1 from Chap. 2 for convenience.

Definition 2.1 Chapter 2

Let's consider an algebraic system $A_1 = \langle A_1; \Omega_1 \rangle$ of the signature Ω_1, and besides it let every element $a_\alpha, \alpha \in \Lambda_1$, of the main set A_1 of the system A_1 in turn be an algebraic system of the signature Ω_2. That is

$a_\alpha = A_2^\alpha = \langle A_2^\alpha; \Omega_2 \rangle, \alpha \in \Lambda_1$ is an algebraic system of the second level. We continue this process by induction. If the algebraic system $A_k^\alpha = \langle A_k^\alpha; \Omega_k \rangle, \alpha \in \Lambda_k$ is the algebraic system of the fractal level k, and every element $a_\alpha, \alpha \in \Lambda_k$ of the main set A_k^α of the system A_k^α is an algebraic system $a_\alpha = A_{k+1}^\alpha = \langle A_{k+1}^\alpha; \Omega_{k+1} \rangle$ of the signature Ω_{k+1}, of the fractal level $k + 1$, $\alpha \in \Lambda_{k+1}$, then the algebraic system $A_1 = \langle A_1; \Omega_1 \rangle$ is called a quasi-fractal algebraic system. If all signatures $\Omega_k, k = 1, \ldots n, \ldots$, are equal to each other (coincide), and all algebraic systems $\langle A_k^\alpha; \Omega_k \rangle, \alpha \in \Lambda_k$, are isomorphic to each other, then the algebraic system $A_1 = \langle A_1; \Omega_1 \rangle$ is called a fractal algebraic system of the signature Ω_1.

In Chap. 3, Sect. 3.2, the concept of quasi-fractal homomorphism was formulated. We give it for convenience here. It runs as follows.

Definition 3.2 Chapter 3, Sect. 3.3.

Let $A_1 = \langle A_1; \Omega_1 \rangle$, $B_1 = \langle B_1; \Omega_1 \rangle$ be quasi-fractal algebraic systems such that for each level of a quasi-fractal systems the systems $A_k^\alpha = \langle A_k^\alpha; \Omega_k \rangle$ and $B_k^\alpha = \langle B_k^\alpha; \Omega_k \rangle$ have one the same signature Ω_k. A mapping $f : A_1 = \langle A_1; \Omega_1 \rangle \rightarrow B_1 = \langle B_1; \Omega_1 \rangle$ is called a quasi-fractal homomorphism from a system $A_1 = \langle A_1; \Omega_1 \rangle$ into a system $B_1 = \langle B_1; \Omega_1 \rangle$, if the following two conditions are fulfilled:

(1) For every $\alpha \in \Lambda_k$ the equality $f(a_\alpha) = f(b_\alpha)$ holds for some $b_\alpha \in \Lambda_k$, that is f maps the elements of the quasi-fractal algebraic system $A_1 = \langle A_1; \Omega_1 \rangle$ of the level $k + 1$ into elements of the same level $k + 1$ of the quasi-fractal algebraic system $B_1 = \langle B_1; \Omega_1 \rangle$;

(2) f saves all operations and predicates of the signature Ω_k for all levels k of the quasi-fractal. That is:

$f\left(F_\xi(x_1, \ldots, x_{m_\xi})\right) = \left(f\left(F_\xi\right)\right)(f(x_1), \ldots, f(x_{m_\xi}))$ for every operation $F_\xi \in \Omega_k$ and every $x_1, \ldots, x_{m_\xi} \in A_k^\alpha$

and

$P_\eta\left((x_1, \ldots, x_{n_\eta})\right) \Leftrightarrow \left(f\left(P_\eta\right)\right)(f(x_1), \ldots, f(x_{n_\eta}))$ for every predicate $P_\eta \in \Omega_k$ and every $x_1, \ldots, x_{n_\eta} \in A_k^\alpha$

7.6.2 Random Quasi-fractal Scales

The aim of this section is to construct a notion of a quasi-fractal scale. It is based on the main constructions from Chap. 4. Namely, in Chap. 4, the general concept of a scale and a quasi-fractal scale was formulated. Let's remained it for convenience.

Definition 4.1 [22]
   Let $\langle A, \Omega \rangle$ и $\langle B, \Omega' \rangle$ be models. An ordered pair

$$f = \langle f_1, f_2 \rangle$$

is called a homomorphism from $\langle A, \Omega \rangle$ into $\langle B, \Omega' \rangle$, if f_1 is a map from A into B, $f_1 : A \to B, f_2$ is a map from Ω into $\Omega', f_2 : \Omega \to \Omega'$, and additionally for any predicate P from Ω and any elements $a_1, a_2, \ldots, a_k$ from A
   $P(a_1, a_2, \ldots, a_k) = 1$ implies $f_2(P)(f_1(a_1), f_1(a_2), \ldots, f_1(a_k)) = 1$ where the predicates P and $f_2(P)$ are of the same rank.

Definition 4.2 Chapter 4. Let $\langle A, \Omega \rangle$ be a model and $\langle B, \Omega' \rangle$, where $B \subseteq R$, be a numerical model. An ordered triple $\langle A, B, f \rangle$, where $f = \langle f_1, f_2 \rangle$ is a homomorphism from $\langle A, \Omega \rangle$ into $\langle B, \Omega' \rangle$, is called a scale.
   Thus, measurement on a scale is the ordering (partial ordering) of a set in accordance with some property (measured on a given scale) and permissible scale transformations, where the notion of a permissible transformation runs as follows.

Definition 4.3 Chapter 4
   A homomorphism $\varphi : \langle R, \Omega', \le \rangle \to \langle R, \Omega', \le \rangle$, where $\le$ is the relation "less than or equal to" on R, is called an admissible transformation of the scale $\langle A, B, f \rangle$, if for an arbitrary fixed homomorphism $f_0 : \langle A, \Omega \rangle \to \langle B, \Omega' \rangle$ the following diagram

$$< A, \Omega > \overset{f_0}{\to} \langle B, \Omega' \rangle \overset{i}{\to} \langle R, \Omega', \le \rangle$$

$$f \searrow \qquad\qquad \varphi \nearrow \qquad\qquad (7.2)$$

$$\langle B, \Omega' \rangle \quad \overset{i}{\to} \quad \langle R, \Omega', \le \rangle$$

is commutative, that is $\varphi i f_0 = i f$:

is commutative, that is $\varphi i f_0 = i f$:

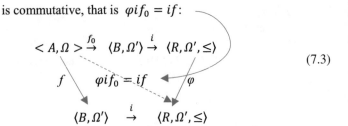

$$(7.3)$$

where $i = \langle i_1, i_2 \rangle$ is a natural embedding of $\langle B, \Omega' \rangle$ into $\langle R, \Omega', \leq \rangle$, that is

$i_1 : B \rightarrow R$ is a natural embedding and $i_2 : \Omega' \rightarrow \Omega' \cup \{\leq\}$ is a natural embedding.

The set of all admissible transformations with respect to the composition operation of the scale $\langle A, B, f \rangle$ forms a group $\langle \Phi, \circ \rangle$ of all admissible transformations of the scale $\langle A, B, f \rangle$, where $\circ$ is the composition operation. An admissible transformation $\varphi \in \Phi$ of the scale $\langle A, B, f \rangle$ convert the scale $\langle A, B, f \rangle$ into its equivalent scale $\langle A, B, \varphi \circ f \rangle$.

The class of equivalent scales is called the type of measurement scale.

Now let's consider some examples.

Examples

1. **Probability space as a scale**. Let $\langle \Omega, \Sigma, p \rangle$-be a probability space. We shall remind the definition of a probability space for convenience. Let Ω be a non-empty set, and $P(\Omega)$ be a set of all subsets of the set Ω. Let's consider a subalgebra

$\langle \Sigma, \cup, \cap, \backslash, \prime \rangle$ of an algebra $\langle P(\Omega), \cup, \cap, \backslash, \prime \rangle$ where $P(\Omega)$ is a set of all subsets of the set Ω. So, the main set Σ of this algebra is closed under all main operations

$\cup, \cap, \backslash, \prime$, that is, for any $A, B \in \Sigma$ the following inclusions: $A \cup B \in \Sigma$, $A \cap B \in \Sigma$, $A \backslash B \in \Sigma$, $A' \in \Sigma$ are true. Elements of the set Ω are called elementary outcomes. Elements of the algebra $\langle \Sigma, \cup, \cap, \backslash, \prime \rangle$, or that is the same, elements of the main set Σ of this algebra are called elementary events. Each of algebras $\langle \Sigma, \cup, \cap, \backslash, \prime \rangle$ is called an algebra of elementary events or an algebra of subsets. Let $\langle \Sigma, \cup, \cap, \backslash, \prime \rangle$ be closed under countable intersections that is if $A_i \in \Sigma$, where $i \in N$, N is a countable set, then $\underset{i \in N}{\cap} A_i \in \Sigma$.

Definition 7.25 An algebra of elementary events is called σ-algebra if it is closed under countable intersections.

Definition 7.26 Let every set A, belonging to some σ-algebra $\langle \Sigma, \cup, \cap, \backslash, \prime \rangle$ of elementary events Ω, put into accordance a real number $p(A)$ such that $0 \leq p(A) \leq 1$, that is a function $p : \Sigma \rightarrow [0, 1]$ is given. Function p is called a probability measure or simply probability if the following conditions or axioms take place:

(1) Countable additivity's condition: the measure of the union of a countable number of pairwise disjoint events is equal to the sum of their measures, that is $p(\cup_{i \in N} A_i) = \sum\limits_{i \in N} p(A_i)$, if $A_i \cap A_j = \emptyset$ for $i, j \in N, i \neq j$.

(2) Normalization condition: the probability of a true event is 1, i.e. $p(\Omega) = 1$. So, we have the diagrams:

$$< \Omega, \Sigma, p > \xrightarrow{f_0} \langle [0,1], p \rangle \xrightarrow{i} \langle R, p, \leq \rangle$$

$$\qquad\qquad p \searrow \qquad\qquad\qquad \swarrow \varphi \qquad\qquad\qquad (7.4)$$

$$\langle [0,1], p \rangle \xrightarrow{i} \langle R, p, \leq \rangle$$

is commutative,

that is $\varphi i f_0 = i p$:

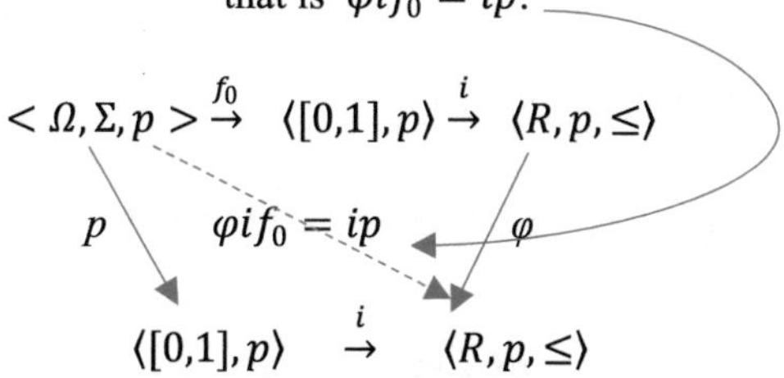

$$< \Omega, \Sigma, p > \xrightarrow{f_0} \langle [0,1], p \rangle \xrightarrow{i} \langle R, p, \leq \rangle$$

$$p \searrow \quad \varphi i f_0 = i p \quad \leftarrow \varphi \qquad\qquad (7.5)$$

$$\langle [0,1], p \rangle \xrightarrow{i} \langle R, p, \leq \rangle$$

Let's give some patterns to prove our approach in Example 1. In [23] calculations in linear probability space are performed and methods to transform data into and out of linear probability space are described.

The following question now arises:

How one can to connect graphs and scales?

To answer this question let's consider probability space $\langle \Omega_n, \Sigma_n, p_n \rangle$ from Sect. 7.3.1. Formal Description of the Probability Space Model as a space in which $\Omega_n = \{ \Gamma_n = \langle V_n, E_n = \{ (u_{v_i}, u_{v_j}) | i, j = 1, \ldots, n \} \rangle \}$, $\Sigma_n = 2^{\Omega_n}$ is the set of all subsets of the set Ω_n, $p_n(\Gamma_n) = p_n^{E_n} (1 - p_n)^{C_n^2 - |E_n|}$.

This construction helps us to connect graphs and scales. Let's embed (or replace) $\langle R, \Omega', \leq \rangle$ in the diagram (7.3) into (on) the metric space. Or let's analyze the diagram (7.3) under assumption that $\langle B, \Omega' \rangle$ is a metric space.

Definition 7.27 Let $A = \langle A, \Omega \rangle = A_1 = \langle A_1; \Omega_1 \rangle$ be a quasi-fractal model and $B = \langle B, \Omega' \rangle = \langle [0, 1], p \rangle = A_1 = \langle A_1; \Omega_1 \rangle$, where $B \subseteq R$, and p is a probability measure, be a numerical quasi-fractal model. Let's remember that in this notation $A_1 = \langle A_1; \Omega_1 \rangle$ plays the role of a variable to denote a quasi-fractal algebraic system. An ordered triple $\langle A, B, f \rangle = C_1 = \langle C_1; \Omega_1 \rangle$, where $f = \langle f_1, f_2 \rangle$ is a quasi—homomorphism from $A = \langle A, \Omega \rangle = A_1 = \langle A_1; \Omega_1 \rangle$ into $B = \langle B, \Omega' \rangle = A_1 = \langle A_1; \Omega_1 \rangle$, is called a random quasi-fractal scale.

Definition 7.28 A homomorphfism $\varphi : R = \langle R, \Omega', \leq \rangle = A_1 = \langle A_1; \Omega_1 \rangle \to R = \langle R, \Omega', \leq \rangle = A_1 = \langle A_1; \Omega_1 \rangle$, where $\leq$ is the relation "equal or less" on the quasi-fractal R, (that is on the each level of the quasi-fractal R) is called an admissible transformation of the random quasi-fractal scale $\langle A, B, f \rangle$, if for every fixed homomorphism

$f_0 : A = \langle A, \Omega \rangle = A_1 = \langle A_1; \Omega_1 \rangle \to B = \langle B, \Omega' \rangle = A_1 = \langle A_1; \Omega_1 \rangle$ the next diagram

$A = \langle A, \Omega \rangle = A_1 = \langle A_1; \Omega_1 \rangle \overset{f_0}{\to} \langle [0,1], p \rangle = A_1 = \langle A_1; \Omega_1 \rangle \overset{i}{\to} \langle R, \Omega', \leq \rangle = A_1 = \langle A_1; \Omega_1 \rangle$

f

$B = \langle B, \Omega' \rangle = \langle [0,1], p \rangle = A_1 = \langle A_1; \Omega_1 \rangle \overset{i}{\to} \langle R, \Omega', \leq \rangle =$
$A_1 = \langle A_1; \Omega_1 \rangle$

is commutative one, that is $\varphi i f_0 = i f$:

$A = \langle A, \Omega \rangle = A_1 = \langle A_1; \Omega_1 \rangle \overset{f_0}{\to} \langle [0,1], p \rangle = A_1 = \langle A_1; \Omega_1 \rangle \overset{i}{\to} \langle R, \Omega', \leq \rangle =$
$A_1 = \langle A_1; \Omega_1 \rangle$

$f \qquad \varphi i f_0 = i f \qquad \varphi$

$B = \langle B, \Omega' \rangle = \langle [0,1], p \rangle = A_1 = \langle A_1; \Omega_1 \rangle \overset{i}{\to} \langle R, \Omega', \leq \rangle = A_1 =$
$\langle A_1; \Omega_1 \rangle$

here $i = \langle i_1, i_2 \rangle$ is a natural embedding $\langle B, \Omega' \rangle = \langle [0, 1], p \rangle = A_1 = \langle A_1; \Omega_1 \rangle$ into $\langle R, \Omega', \leq \rangle$, that is $i_1 : B \to R$—is a natural embedding and $i_2 : \Omega' \to \Omega' \cup \{\leq\}$—is a natural embedding (that is, the commutativity of the diagram takes place at each level of all quasi-fractals involved in it).

It follows from Definition 7.28 that a quasi-fractal scale can be constructed in such a way that it contains all the scales corresponding to Chap. 4, Table 4.1 of the classification of scales and their generalizations—multidimensional, matrix, and lattice scales.

References

1. Home page. www.differencebetween.com/difference-between-structure-and-vs infrastructure/
2. Home page. https://en.wikipedia.org/wiki/Infrastructure

3. Raygorodsky, A.M.: Models of random graphs and their applications. Proc. MIPT **2**(4), 130–140 (2010)
4. Serdyukova, N., Serdyukov, V.: Algebraic Formalization of Smart Systems. Theory and Practice, Smart Innovation, Systems and Technologies, SIST, vol. 91, Springer Nature, Switzerland (2018)
5. Home page. https://en.wikipedia.org/wiki/Random_graph
6. Bollobás, B.: Random Graphs, 2nd edn. Cambridge University Press, Cambridge (2001)
7. Raygorodsky, A.M.: Models of Random graphs. Moscow Publishing House, Moscow (2011)
8. Home page. https://en.wikipedia.org/wiki/Erdős–Rényi_model
9. Frieze, A., Karonski, M.: Introduction to Random Graphs. Cambridge University Press, Cambridge (2016)
10. Serdyukova, N.A.: On generalizations of purities. Algebra Logic **30**(4), 432–456 (1991)
11. Zolotarev, V.M.: Probabilistic Metrics. Prob. Theory Appl. **18**(2), 264–287 (1983)
12. Shelah, S., Spencer, J.: Zero-one laws for sparse random graphs. J. Am. Mathem. Soc. **1**(1), 97–115 (1988)
13. Semenov A.S.: Fractal developing architectures, Large systems management, Special issue 30.1 Network models in management, Publishing House: V.A. Trapeznikov Institute of management problems of the Russian Academy of Sciences, Moscow, pp. 91–103 (2010)
14. Danilov, V.I.: Lectures on Fixed Points. Russian Economic School, Moscow (2006). (in Russian)
15. Erdös, P., Rényi, A.: On random graphs I. Publ. Math. Debrecen **6**, 290–297 (1959)
16. Erdös, P., Rényi, A.: On the evolution of random graphs. Publ. Math. Inst. Hungar. Acad. Sci. **5**, 17–61 (1960)
17. Erdős, P., Rényi, A.: On the evolution of random graphs. Bull. Inst. Int. Statist. Tokyo **38**, 343–347 (1961)
18. Chein, M.; Mugnier, M.-L.: Graph-Based Knowledge Representation: Computational Foundations of Conceptual Graphs (2009). Home page http://www.lirmm.fr/gbkrbook/
19. Home page. https://en.wikipedia.org/wiki/Path_(graph_theory)
20. Kuznetsov, O.P., Adelson-Velsky, G.M.: Discrete mathematics for the engineer. Energia, Moscow (1980). (in Russian)
21. Home page. https://en.wikipedia.org/wiki/Rado_graph
22. Serdyukova, N.A.: Optimization of Tax System of Russia. Budget and Treasury Academy, Rostov State Economic University, Moscow, Parts I and II (2002). (in Russian)
23. Rembold, C.M.: Statistical testing in a Linear Probability Space, Home page. https://www.res earchgate.net/publication/332169679

Chapter 8
Cayley Graphs and Probability Isomorphic Groups

"There are also a number of seemingly
unrelated problems from physics,
communication engineering, statistics and so on,
that lead us to consider probabilistic relations
in algebraic structures not equivalent to real line."
Ulf Grenander, Probabilities on Algebraic Structures.

Abstract In this chapter the following basic questions are considered:

- Cayley diagrams, Cayley complexes and Cayley groups graphs, the relationship between them through the basic concepts of category theory,
- Van Kampen diagrams, relationship of Cayley diagrams and Van Kampen diagrams,
- a brief overview of graph study methods,
- probabilistic properties of graphs,
- the use of first-order logic in the study of graphs,
- category concept, examples, role in mathematics and use in systems theory,
- the relationship between groups and graphs through category theory,
- the possibility of transferring the theory of P—purities of algebraic systems to graphs,
- probability-isomorphic finite quasi-fractal groups as analogues of giant components.

Keywords Probability-isomorphic groups · Category concept · Giant component

8.1 Introduction

In this chapter we define the concept of probability-isomorphic groups. We introduce this concept using Erdos–Renyi algorithm on finite graphs. To define probability isomorphic groups, we need the concept of a Cayley graph of a group. Therefore, in this chapter we shall consider the following issues:

© Springer Nature Switzerland AG 2021
N. A. Serdyukova and V. I. Serdyukov, *Algebraic Identification of Smart Systems*, Intelligent Systems Reference Library 191,
https://doi.org/10.1007/978-3-030-54470-6_8

- Cayley diagrams, Cayley complexes and Cayley groups graphs, the relationship between them through the basic concepts of category theory,
- Van Kampen diagrams,
- relationship of Cayley diagrams and Van Kampen diagrams,
- a brief overview of graph study methods,
- probabilistic properties of graphs,
- the use of first-order logic in the study of graphs,
- category concept, examples, role in mathematics and use in systems theory,
- the relationship between groups and graphs through category theory,
- the possibility of transferring the theory of P—purities of algebraic systems to graphs,
- probability-isomorphic finite quasi-fractal groups as analogues of giant components [1].

In this chapter, the concept of the Cayley graph of a group and the questions related to this concept are important:—Cayley diagrams, Cayley complexes, Van Kampen diagrams, Van Kampen probability diagrams, the connection of Cayley diagrams and Van Kampen diagrams, the connection of the Cayley graph of a group and Van Kampen diagrams through the basic concepts of category theory, category concept, examples, their role in mathematics and their using in systems theory, the relationship between groups and graphs through category theory. The notion of probability-isomorphic groups is introduced. It is an analogue of giant component [1]. We shall consider some notions of Category Theory because many algebraic notions are in deep connection with it: many algebraic objects, such as a group, a monoid, a field of fractions of a domain of integrity, etc. can be represented as a category [2]. On the other hand, each category C can be considered as a graph [2]. So, the study of these issues has shown that the problem of identifying a system is extremely important in the study of mathematical objects. Let's begin with significant necessary preliminary notions from [2].

8.1.1 Preliminary Notions. History of the Question

We shall start by examining various approaches to the study of graph theory.

There are many different approaches to study Graph Theory. Let us dwell upon some of them:

- these are first of all combinatorial methods of analyses of graphs—given binary relation on the set of vertices of a graph,
- algebraic methods (especially in enumeration tasks), and foremost methods of algebraic systems, groups theory (especially of substitution groups), semigroups theory, matrix theory methods,
- an elementary theory of category theory is considered in [2, p. 31],
- model's theory methods [3],

- one can highlight methods of theory of categories, topological and geometric methods [2, 4],
- probabilistic methods are used to study random graphs,
- categories theory methods.

First of all, we want to clear up the connection between Cayley diagrams, Cayley complexes and Cayley graphs of groups, Van Kampen diagrams. Let's begin with connections between Categories Theory and Graph Theory.

8.1.2 Categories Theory Methods in Graph Theory. Approach to Graph Definition Through Category Theory

In this section we shall use wonderful book by Saunders Mac Lane [2]. Mac Lane used a very clear and accessible description of the concepts of category theory in [2], which we will use.

Definition 8.1 [2] A category is a monoid for the product in the general sense described in the following way.

A metagraph consists of objects $a, b, c, \ldots$, arrows $f, g, h, \ldots$, and two operations as follows:

Domain, which assigns to each arrow f an object $a = dom f$; codomain which assigns to each arrow f an object $b = cod f$. These operations on f are best indicated by displaying f as an actual arrow starting at its domain and ending at its codomain. A metacategory is a metagraph with two additional operations: identity, which assigns to each object a an arrow $id_a = 1_a : a \rightarrow a$; composition which assigns to each pair g, f of arrows with $dom g = cod f$ an arrow $g \circ f$ called their composite, with $g \circ f : dom f \rightarrow cod g$. This operation is pictured by the diagram

These operations in a metacategory are subjects to the two following axioms:
Associativity. For given objects and arrows in the configuration

$$a \xrightarrow{f} b \xrightarrow{g} c \xrightarrow{k} d$$

One always has the equality

$$k \circ (g \circ f) = (k \circ g) \circ f \tag{8.1}$$

This equation is represented pictorially by the statement that the following diagram is commutative:

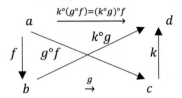

Unit law. For all arrows $f : a \to b$ and $g : b \to c$ composition with the identity arrow 1_b gives

$$1_b^\circ f = f \text{ and } g^\circ 1_b = g \tag{8.2}$$

This axiom asserts that the identity arrow 1_b of each object b acts as an identity for the operation of composition, whenever it makes sense. The arrows of any metacategory are often called its morphisms.

A metacategory is to be any interpretation which satisfies all these axioms. An example is a metacategory of sets, which has objects all sets and arrows all functions, with the usual identity functions and the usual composition of functions. Function means a function with specified domain and specified codomain. A function $f :$ $X \to Y$ consists of a set X, its domain, a set Y, its codomain, and a rule, which assigns to each element $x \in X$, an element $fx \in Y$. A metacategory of all sets has as object, all sets, as arrows, all functions with the usual composition. The metacategory of all groups is described in the following lay: objects are all groups, arrows are all homomorphisms of groups. The arrows of any metacategory are called its morphisms. Since the objects of a metacategory corresponds exactly to its identity arrows, it is technically possible to dispence altogether with the objects and deal only with arrows. The data for an arrows—only metacategory C consist of arrows, certain odered pairs $\langle g, f \rangle$ called the composable pairs of arrows, and an operation assigning to each composable pair g, f an arrow $g \cdot f$ called their composite. With these data one defines an identity of C to be an arrow u such that $f \cdot u = f$ whenever the composite $f \cdot u$ is defined and $u \cdot g = g$ whenever the composite $u \cdot g$ is defined. The data are then required to satisfy the following three axioms:

(1) The composite $(k \cdot g) \cdot f$ is defined if and only if the composite $k \cdot (g \cdot f)$ is defined. When either is defined, they are equal, and this triple composite is written as $k \cdot g \cdot f$.
(2) The triple composite $k \cdot g \cdot f$ is defined whenever both composites $k \cdot g$ and $g \cdot f$ are defined.
(3) For each arrow g of C there exist identity arrows u and u' of C such that

$u' \cdot g$ and $g \cdot u'$ are defined.

So a category as distinguished from a metacategory will mean any interpretation of the category axioms within set theory. The details run as follows.

Definition 8.2 [2] A directed graph (also called a diagram scheme) is a set O of objects, and a set A of arrows, and two functions

$$A \overset{dom}{\to} O, A \underset{cod}{\to} O$$

In this graph the set of composable pairs of arrows is the set

$$A \times_O A = \{\langle g, f \rangle | g, f \in A, dom\, g = cod\, f\},$$

is called the product over O. A graph is small if both O and A are small sets, that is both O and A are members of the big enough set U, the universe.

Definition 8.3 [2] A category is a graph with two additional functions

$$\begin{array}{cc} O \overset{id}{\to} A, & A \times_O A \overset{\circ}{\to} A, \\ c \mapsto id_c, & \langle g, f \rangle \mapsto g \circ f, \end{array} \tag{8.3}$$

called identity and composition and also written as gf, such that

$$dom(ida) = a = cod(ida), dom(g \circ f) = dom f, cod(g \circ f) = cod g \tag{8.4}$$

for all objects $a \in O$ and all composable pairs of arrows $g, f \in A \times_O A$, and such that the associativity and unit axioms (1)–(3) hold.

Examples [2]

1. Discreate categories. A category is descreate when every arrow is an identity. Every set X is the set of objects of a discrete category, just add one identity arrow $x \to x$ for each $x \in X$, and every discreate category is determined by its set of objects. Thus, discrete categories are sets.
2. Monoids. A monoid is a category with one object. So each monoid is determined by the set of all its arrows, by the identity arrow, and by the rule for the composition of arrows. For any category C any object $a \in C$, the set $hom(a, a)$ of all arrows $a \to a$ is a monoid.
3. Groups. A group is a category with one object in which every arrow has a two-sided inverse under composition.

So [2], every category C determines a graph UC with the same objects and arrows, forgetting which arrows are composites and which are identities. Every functor $F :$ $C \to C'$ is also a morphism $UF : UC \to UC'$ between the corresponding graphs.

So the category of groups is a graph. Besides it one can consider the category of all small graphs. Let's explain the word "small". In some choices of foundations, one says small set in order to amplify that one really means a set and not a proper class. However, in other choices of foundations, such as Grothendieck universes, there exist both "small sets" (sets that belong to the universe) and "large sets" (sets that do not, such as the universe itself).

Definition 8.4 A category is **small** if it has a small set of objects and a small set of morphisms.

Let $Grph$ be the category of all small graphs, the word "small" we understand such as we mentioned, and as in [2], that is we add the assumption: the existence of univers.

There exists a deep connection between groups and graphs. Matumoto [5] proved that for any graph Γ there exists a group G whose outer automorphism group is isomorphic to the group of automorphisms of Γ. So, every group can be realized as the group of automorphisms of some graph. The category Grp of all small groups can be embedded into the category of all small graphs $Grph$ through Cayley diagrams, Cayley complexes and Cayley graphs of groups. Conversely, there are a number of papers in which it is shown that there exist embeddings (almost complete) of the category of graphs into the category of groups [6]. A functor $F : Grph \to Grp$ is constructed in it such that F is faithful and "almost " full, that is that every nontrivial group homomorphism $FX \to FY$ is a composition of an inner automorphism of FY and a homomorphism of the form Ff, for a unique map of graphs $f : X \to Y$.

Some remarks

The notion of monoid (a semigroup with identity) plays a central role in category theory. A monoid M may be described as a set M together with two functions $\mu : M \times M \to M, \eta : 1 \to M$ such that the following diagrams in μ and η commute:

$$
\begin{array}{ccc}
M \times M \times M & \xrightarrow{1 \times \mu} & M \times M \\
{\scriptstyle \mu \times 1}\Big\downarrow & & \Big\downarrow{\scriptstyle \mu} \\
M \times M & \xrightarrow{\mu} & M
\end{array}
\qquad
\begin{array}{ccccc}
1 \times M & \xrightarrow{\eta \times 1} & M \times M & \xleftarrow{1 \times \eta} & M \times 1 \\
{\scriptstyle \lambda}\Big\downarrow & & \Big\downarrow{\scriptstyle \mu} & & \Big\downarrow{\scriptstyle \rho} \\
M & = & M & = & M
\end{array}
$$

Here 1 in $1 \times \mu$ is the identity function $M \to M$, and 1 in $1 \times M$ is the one-point set
$1 = \{0\}$, while λ, ρ are the bijections:

$$1 \times X \xrightarrow{\lambda} X \xleftarrow{\rho} X \times 1$$

To say that these diagrams commute means that the following composites are equal:

$$\mu^\circ(1 \times \mu) = \mu^\circ(\mu \times 1), \mu^\circ(\eta \times 1) = \lambda, \mu^\circ(1 \times \eta) = \rho.$$

These diagrams may be rewritten with elements writing the function μ (say) as a product $\mu(x, y) = xy$ for $x, y \in M$ and replacing the function η on one-point set $1 = \{0\}$ by its only value, an element $\eta(0) = u \in M$.

The diagrams above then become.

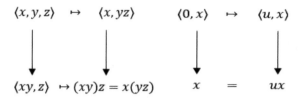

They are exactly the familiar axioms on a monoid."

A groupoid is a category in which every arrow is invertible.

8.2 Cayley Diagrams, Cayley Complexes and Cayley Groups Graphs

We shall use the definitions of a complex, a skeleton of a graph, a one-dimensional skeleton of a graph in this section and clear up the difference between these notions.

8.2.1 Cayley Diagrams

Representation of a group as a network consisting of directed segments or edges, where the vertices of the edges correspond to the elements and the segments to multiplication by the generating elements of a groups and the inverse elements of a group, was introduced by Cayley in the nineteenth century [7]. We shall need the concepts of "group alphabet" and "group word" for further. Let's define them following [8].

Definition 8.5 [8].

Let A be non-empty set of symbols or letters. Let's designate by A^{-1} the set of symbols a^{-1}, where $a \in A$, and herewith it is assumed that there are no symbols a^{-1} in A, where $a \in A$. So $A^{-1} = \{a^{-1} | a \in A\}$. Besides it, we assume that $(a^{-1})^{-1} = a$. The set $A \cup A^{-1}$ is called the group alphabet, and its elements are called letters. Any finite sequence $x_1 \ldots x_n$, where $x_i \in A \cup A^{-1}, i = 1, \ldots, n$, is called a word or a group word in the alphabet $A \cup A^{-1}$. The number n is called the word length of $X = x_1 \ldots x_n$ and it is designated by $|X|$. We shall consider a word of zero length - an empty word. The word $X = x_1 \ldots x_n$ is called reducible if for some $i, i = 1, \ldots, n$, letters x_i and x_{i+1} are mutually inverse. If one crosses the mutually inverse letters x_i and x_{i+1} then a word of a length $n - 2$ remains. In such case we say that it is a result of reducing of X. If after several reducing, we have unreducible word $X\prime$ from X, then $X\prime$ is called the result of full reducing of X. Let's define multiplication of two irreducible words X and Y in the alphabet $A \cup A^{-1}$ as the word XY if reducing is impossible at the junction of the words X and Y. If one can to reduce the word XY then the result of multiplication of X and Y is defined as the result of full reducing of the word XY.

The following theorem holds.

Theorem 8.6 [8].

The set $F(A)$ of all irreducible words in the alphabet $A \cup A^{-1}$ is a group with respect to the word multiplication operation introduced in Definition 8.5.

Let's now give a geometric interpretation of the introduced concepts, following [7]

Example 8.7 Let's consider for example cyclic group Z_3 of the order 3, and construct its Cayley diagram. If x is any element of cyclic group of order 3, then every word, representing the element x can be presented as movement along a graph of a triangle form with vertices marked with the symbols $a, a^2, I = a^3$ (Fig. 8.1).

The movement in the direction indicated by the arrow corresponds to the multiplication on the right by the generating element a of the group. Movement in the direction opposite to the arrow corresponds to the multiplication on the right by the element a^{-1}, inverse to a. So, the word corresponding to the element X is represented as a sequence of directions when moving along a path in the graph. To each word X corresponds a certain sequence of movements along successively connected directed segments—oriented edges of the graph, and, conversely, any path along the oriented edges of the graph of the group starting from the vertex I designating the neutral element of the group, corresponds to a certain word.

8.2.2 Cayley Diagram Construction of an Arbitrary Group

Let's now give an algorithm for constructing a Cayley diagram of an arbitrary group. The same construction as in previous example can be applied to an arbitrary group defined by a set of generators and a set of defining relations.

Herewith:

- the graph will have as many vertices as there are elements in the main set of the group,

Fig. 8.1 Geometric interpretation of words multiplication in a group. Cayley diagram of a group Z_3

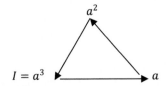

- at each vertex of the graph two edges converge, the movement along one of which corresponds to the multiplication on the right by the generating element a, and the movement along the other corresponds to the multiplication to the right by the element a^{-1}, if the group has one generating element a. If a group has two generators a and b, then at each vertex of the graph four edges converge, corresponding to multiplication on the right by the elements a, a^{-1}, b, b^{-1}. If a group has n generators $a_1, \ldots, a_n$, then at each vertex of the graph $2n$ edges converge corresponding to multiplication on the right by elements $a_1, \ldots, a_n, a_1^{-1}, \ldots, a_n^{-1}$,
- each word in the group alphabet one-to-one corresponds to a path in a graph—a sequence of oriented edges of a graph, and, vice versa,
- the multiplication of two elements of the group corresponds to the passage of the path of the graph, composed of two consecutive paths. The product $XY = Z$ of the elements X and Y can be interpreted as a path in the graph, which is constructed as follows. Let's write the words X and Y as words off generators and their inverse elements. Then coming from the vertex corresponding to the neutral element of group I, let's go along the path in the graph corresponding to the word X. The endpoint of this path corresponds to element X. Now, taking the vertex of the graph corresponding to X as the starting point, go through the path corresponding to the element Y. This path will end at the vertex of the graph that corresponds to the element $XY = Z$, regardless of which words are used to represent the elements X and Y. Any word W that corresponds the element I defines a closed path, regardless of which point is taken as the starting point,
- the graph of the group is connected, that is, there is a path from any of its vertices to any of its other vertices.

8.2.3 Cayley Group Diagram Regard to the Subset of its Elements

We shall follow now [9, p. 174]. Cayley showed that each group is isomorphic to a permutation group. Indeed, the right regular representation Δ maps the group G isomorphically onto some permutation group of its elements: to each element $g \in G$ there corresponds a permutation $g\Delta : h \to hg$. Let's represent the set $g\Delta = \{(h, hg) | h \in G\}$ in the form of a graph with the set of vertices $V = G$ and with oriented edges from h to hg. If we consider several different permutations of $g\Delta$ at once, then we can distinguish them by coloring the edges belonging to different $g\Delta$ in different colors. This is the Cayley group G diagram regarding to the set $X \subseteq G$ of the elements $g \in G$ under consideration. This diagram is connected if and only if the set of elements of X generates the group G.

8.2.4 Cayley Group Complex

Now let's slightly modify the notation and now suppose that G is given as a quotient group of a free group F with basis X. For $h \in G$ and $\omega \in F$, we denote by $h\omega$ the product in G of the element h and the image of the word ω. After adding inverse edges, this gives the Cayley diagram of G corresponding to the one-dimensional skeleton of the Cayley complex.

Definition 8.8 Let $F(X)$ be a free group Then a word ω in $F(X)$ is said to be cyclically reduced if
and only if every cyclic permutation of the word ω is reduced.

Definition 8.9 Let $G = (X; R)$, where R is the set of cyclically reduced words in the free group F. The Cayley complex $C = (X; R)$ of the representation $G = (X; R)$. is constructed as follows. For the set of vertices, we choose $V = G$, for the set of edges we choose the Cartesian product $G \times L$, where $L = X \cup X^{-1}$, the set of faces is $G \times (R \cup R^{-1})$. The edge (g, y) for $g \in G$, $y \in L$ goes by definition from g to gy, the inverse edge to it is (gy, y^{-1}). The Cayley complex is equipped with a marking function, which is defined as follows. Let's associate a label $e\varphi = y$ to each edge $e = (g, y)$. Then we have $(e^{-1})\varphi = (e\varphi)^{-1}$. After that we continue the function φ multiplicatively: the word $\rho\varphi = (e_1\varphi) \ldots (e_n\varphi)$ is associated as a label to a path $\rho = e_1 \ldots e_n$. It turns out that $\rho\varphi$ is a reduced word if and only if ρ is a reduced path. For any vertex v, the function φ establishes a one-to-one correspondence between the set of paths starting with v and the set of all words. We shall use the notation $\rho\varphi$ for elements from F and for elements from G defined by $\rho\varphi$. Let's note that if ρ begins in v, then ρ ends in $v(\rho\varphi)$, ρ is a loop if and only if $\rho\varphi$ is contained in the normal closure of the set R in F. For $g \in G$ and $s \in R \cup R^{-1}$, the face D is determined by the boundary ∂D of the cycle defined by the loop ρ with the origin g and the label $\rho\varphi = s$.

Let's note that the terms graph and one-dimensional complex are often used as synonyms.

8.2.5 Cayley Planar Complexes

Let's dwell upon the concepts of one-dimensional and two-dimensional complexes. Following [9] we define the one-dimensional complex C as the combination of two sets V and E and three maps $\alpha : E \to V, \omega : E \to V, \eta_1 : E \to E$. We call elements of the set V vertices or points; elements of the set E are called edges. Also, we call $\alpha(e)$ the beginning of the edge $e \in E$, and $\omega(e)$ the end of this edge, the edge $\eta_1(e) = e^{-1}$ will be called inverse to e or opposite oriented. Herewith η_1 should be an involution without fixed elements, and the edge e^{-1} must go from $\omega(e)$ to e. In essence, by this an undirected graph is defined, since it is required that the complex C should contain the opposite one with each edge. Under an undirected edge one means a pair (e, e^{-1}). A path in C is a finite sequence of edges which is designated by $\rho = e_1 \dots e_n, n \geq 1$, at that for $1 \leq i \leq n$ an edge e_{i+1} starts, where e_i ends that is $\alpha(e_{i+1}) = \omega(e_i)$. If these two points coincide then ρ is called a loop. It is convenient [9] not to introduce a single empty path but to define for every vertex v a path 1_v without edges starting and ending in v. The inverse path to ρ is a path $\rho^{-1} = e_1^{-1} \dots e_n^{-1}$. If $\rho = e_1 \dots e_n$ is a loop, then every cyclic permutation $\rho' = p_i \dots p_n p_1 \dots p_{i-1}$ of a path ρ also is a loop. The set of all cyclic permutations of the loop ρ is called a cyclic path or a cycle. A path is reduced if it does not contain a path of the form ee^{-1}. A loop or a cycle is called cyclically reduced if it is reduced and $e_1 \neq e_n^{-1}$.

A path is called a simple one if for $i \neq j$ we have $\alpha(e_i) \neq \alpha(e_j)$ and $\omega(e_i) \neq \omega(e_j)$.

The two-dimensional complex C consists of one-dimensional complex C^1 (its one-dimensional skeleton), the set F of two-dimensional cells or faces, and two mappings, ∂ and η_2, defined on F. The map ∂ associates every cell D from F with cyclically reduced cycle ∂D from C^1 which is a boundary of a cell D. The map η_2 associates every face D with another face $\eta_2(D) = D^{-1}$ which is inverse to D. Herewith it is required that the cycle $\partial(D^{-1})$ should be inverse to $\partial(D^{-1})$. The vertex v lies on the face D if it is the beginning of some edge in ∂D. A contour or a boundary path with start at point v is any loop in the cycle ∂D which starts at point v. If the cycle ∂D is a simple one, then the contour for each vertex of the cell D is unique.

8.2.6 Cayley Group Graph as One-Dimensional Cayley Complex

Let us now dwell upon the concept of Cayley graph of a group. The Cayley group graph is constructed by a group with a distinguished system of generators. Suppose that it is given a discrete group G and its system of generators S, that is for every $s \in S$, an inclusion $s^{-1} \in S$ is true. The Cayley group G graph on the system of generators S is a graph whose vertices are the elements of the group and the element $g \in G$ is connected by an edge with exactly those elements that are obtained by multiplying g by an element from S.

Let's now state the definition of the Cayley graph of an arbitrary group, following [10].

Definition 8.10 [10]

Let G be an arbitrary group and $M \subset G$ be a subset of its elements, $e \notin M$, $M^{-1} = \{g^{-1} | g \in M\} = M$, that is $g, g^{-1} \in M$. Cayley group G graph $\Gamma(G, M)$ is undirected graph with the set of vertices G and with the set of edges $U(\Gamma) = \{\{g, h\} | gh^{-1}, hg^{-1} \in M\}$. Graph $\Gamma_{(G,M)}$ will be a complete graph if and only if $M = G \backslash \{e\}$

Let's consider some examples of Cayley groups graphs (Figs. 8.2, 8.3 and 8.4).

Fig. 8.2 Cayley graph of the free product of a second-order cyclic group and a third-order cyclic group $Z_2 * Z_3$

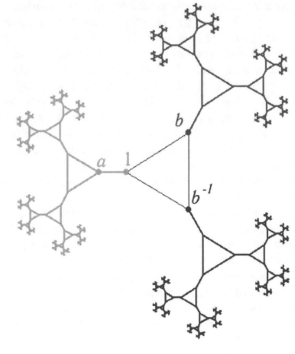

Fig. 8.3 Cayley graph of the Cartesian product of a second-order cyclic group and a third-order cyclic group $Z_2 \times Z_3$

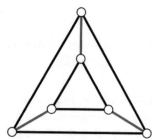

Fig. 8.4 Cayley graph of a free non-Abelian group F_2 with two generators a, b [11]

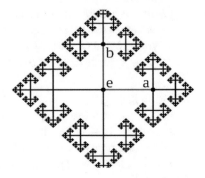

8.2.7 Cayley Group Graph and van Kampen Diagram Connection

Van Kampen diagram is a planar diagram used to represent the fact that a particular word in the generators of a group given by a group presentation represents the identity element in that group.

Definition 8.11 [12]

Van Kampen diagram over the presentation $G = \langle A|R \rangle$ is a planar finite cell complex[1] D, given with a specific embedding $D \subseteq R^2$ with the following additional data and satisfying the following additional properties:

1. The complex D is connected and simply connected.[2]
2. Each edge (one-cell) of D is labelled by an arrow and a letter $a \in A$.
3. Some vertex (zero-cell) which belongs to the topological boundary of $D \subseteq R^2$ is specified as a base-vertex.
4. For each region (two-cell) of D for every vertex the boundary cycle of that region and for each of the two choices of direction (clockwise or counter- clockwise) the label of the boundary cycle of the region read from that vertex and in that direction is a freely reduced word in $F(A)$ that belongs to R_*, where R_* is the symmetrized closure of R, that is, R_* is obtained from R by adding all cyclic permutations of elements of R and of their inverses.

Thus the 1-skeleton of D is a finite connected planar graph Γ embedded in R^2 and the two-cells of D are precisely the bounded complementary regions for this graph.

An abstract diagram is a Van Kampen diagram in which we forget the actual relators associated with the faces, but only remember: the geometry of the diagram, which faces bear the same relator, the orientation, and the beginning points of relators.

Remark The edges of Cayley group graph are not labeled. The Cayley group graph is undirected.

The Cayley group diagram has labeled edges, as is the Van Kampen diagram. We shall need a definition of an edge graph.

Definition 8.12 [13, 14] Let there be given a graph Γ. An edge graph of a graph Γ is a graph $L(\Gamma)$ such that:

– any vertex of the graph $L(\Gamma)$ represents an edge of the graph Γ,
– two vertices of the graph $L(\Gamma)$ are adjacent if and only if their corresponding edges have a common vertex in Γ.

---

[1]Closure—finite $W(CW)$ complex is a type of topological space introduced by J. H. C. White- head for the needs of Homotopy theory. It is made of basic building blocks called cells. The precise definition prescribes how the cells may be topologically glued together. The C stands for "closure-finite", and the W for "weak" topology.

[2]A topological space X is called simply connected if it is path-connected and any loop in X defined by $F : S^1 \rightarrow X$ be contracted to a point.

Fig. 8.5 Graph K_3

Fig. 8.6 Graph $K_{1,3}$

We give the classical Whitney theorem characterizing edge graphs.

Theorem (Whitney) [13, 14] Let Γ and Γ' be connected graphs, such that their edge graphs are isomorphic. Then graphs Γ and Γ' are always isomorphic, except when one of them is K_3 and the other is $K_{1,3}$.

The graph K_3 is a triangle. (The complete graph K_p has $C_p^2 = \frac{p(p-1)}{2}$ edges and is a regular one of a degree $(p-1)$. A regular or homogeneous graph of degree r is a graph with all vertices of the same degree r) (Fig. 8.5).

A bipartite graph or a biograph Γ is a graph such that its set of vertices V can be divided into two sets V_1 and V_2 in a way that each edge of graph Γ connects vertices from different sets. It is said in this case that the edges of the graph Γ connect the sets V_1 and V_2. If herewith the graph Γ contains all edges which connect sets V_1 and V_2, then graph Γ is called a complete bipartite graph. If wherein the number of vertices in V_1 equals to m, and the number of vertices in V_2 equals to n, then the graph Γ is designated by $K_{m,n}$ (Fig. 8.6).

Now let's remember Pontryagin—Kuratovsky theorem, or Kuratovsky theorem in graph theory. It is a theorem in graph theory that gives a necessary and sufficient condition for the planarity of a graph. The theorem states that the graphs K_5 (a complete graph with 5 vertices) and $K_{3,3}$ (a complete bipartite graph with 3 vertices in each fraction) are the only minimal nonplanar graphs.

8.3   Some Other Constructions. $P-$ Purities in Graphs Theory

Further:

– we generalize the concept of P—purities, P—pure exact sequences onto graph category, onto quasi-fractal algebraic systems, onto quasi-fractal graphs. This can be done using the fact that a graph is a semigroup. We now single out groups from the semigroup class and consider those graphs that are groups. The theory of P—purities in the class of groups can be fully applied to them,

- we can represent a category as a graph, so one can spread P—purities theory onto category theory,
- then we embed the graphs in multidimensional Euclidean vector spaces and consider the theory of P—pure graph in multidimensional Euclidean vector spaces.

Let's note that P—pure exact sequences were considered in [15].

8.3.1 A Special Case. Multidimensional Graphs in Euclidean Vector Spaces

Definition 8.13 Flat graph $\Gamma_1 = \left\langle \left\{ V_{\alpha_1}^1 | \alpha_1 \in \Lambda_1 \right\}, \left\{ u_{\alpha_1,\beta_1}^1 | \alpha_1, \beta_1 \in \Lambda_1 \right\} \right\rangle$, where $\left\{ V_{\alpha_1}^1 | \alpha_1 \in \Lambda_1 \right\}$ is the set of vertices of graph Γ_1, $\left\{ u_{\alpha_1,\beta_1}^1 | \alpha_1, \beta_1 \in \Lambda_1 \right\}$ is the set of edges of graph Γ_1, is called a multidimensional Euclidean flat graph (n—dimensional flat Euclidean graph) if its vertices $\left\{ V_{\alpha_1}^1 | \alpha_1 \in \Lambda_1 \right\}$ are the points of Euclidean vector space that is to every point $V_{\alpha_1}^1$, $\alpha_1 \in \Lambda_1$, there corresponds an ordered tuple $\left(x_1^1, \ldots, x_n^1 \right)$, $\alpha_1 \in \Lambda_1$, and the following conditions take place:

(1) to different vertices different n—tuples are corresponded,
(2) there exist $n - 2$—tuple $\left(x_{i_1}, x_{i+1_1}, \ldots, x_{i_1+n-2} \right)$ such, that for all vertices of graph Γ_1 n—tuples which correspond them have equal values, that is for every $V_{\alpha_1}^1$, $V_{\alpha_2}^1$ vertices of Γ_1 if

$$V_{\alpha_1}^1 \mapsto (x_1, \ldots, x_n), V_{\alpha_2}^1 \mapsto (y_1, \ldots, y_n)$$

then $\left(x_{i_1}, x_{i+1_1}, \ldots, x_{i_1+n-2} \right) = \left(y_{i_1}, y_{i+1_1}, \ldots, y_{i_1+n-2} \right)$

Here $u_{\alpha_1,\beta_1}^1 = \left(V_{\alpha_1}^1, V_{\beta_1}^1 \right)$ is an edge which connects the vertices $V_{\alpha_1}^1$, $V_{\beta_1}^1$ in that order.

Now let's normalize all vectors $\left(V_{\alpha_1}^1, V_{\beta_1}^1 \right)$ all edges of graphs Γ_1. We get from here that all the concepts of graph theory carry over to the case when the graph Γ_1 is placed in accordance with Definition 8.13 in a Euclidean n—dimensional vector space.

The metric defined in the multidimensional Euclidean vector space $E_1^n = \langle E_1^n, +, \{\lambda | \lambda \in R\}\rangle$, induces a metric on $\varepsilon \Gamma_1$, where $\varepsilon : \Gamma_1 \to E_1^n$ is the embedding Γ_1 into E_1^n, constructed in definition 8.13.

Now we embed the quasi-fractal graph $\Gamma_1 = \left\langle \left\{ V_{\alpha_1}^1 | \alpha_1 \in \Lambda_1 \right\}, \left\{ u_{\alpha_1,\beta_1}^1 | \alpha_1, \beta_1 \in \Lambda_1 \right\}, Q^1, W^1 \right\rangle = A_1 = \langle A_1; \Omega_1 \rangle$ which is flat in each its level into multidimensional quasi-fractal Euclidean vector space $E_1^n = \langle E_1^n, +, \{\lambda | \lambda \in R\}\rangle = A_1 = \langle A_1; \Omega_1 \rangle$ in the following way: i—the coordinate corresponds to the i—th level of the quasi-fractal and for each i, an embedment is carried out according to the Definition scheme 8.13.

The metric defined in the multidimensional quasi-fractal Euclidean vector space $E_1^n = \langle E_1^n, +, \{\lambda | \lambda \in R\}\rangle = A_{qf} = \langle A_{qf}; \Omega_{qf}\rangle$ induces the metric on $\varepsilon\Gamma_1 = A_1 = \langle A_1; \Omega_1\rangle$, where $\varepsilon : \Gamma_1 = A_1 = \langle A_1; \Omega_1\rangle \to E_1^n = A_{qf} = \langle A_{qf}; \Omega_{qf}\rangle$ is the above embedding $\Gamma_1 = A_1 = \langle A_1; \Omega_1\rangle$ into $E_1^n = A_1 = \langle A_1; \Omega_1\rangle$.

For what follows, we need to establish a connection between groups and graphs from the standpoint of category theory and, in addition to the metrics discussed in Chap. 2, consider the category of metric spaces and give various examples of metrics. Let's do it following [2, 16]. Recall that the category of metric spaces has metric spaces as objects (in the usual sense), and arrows can be entered in different ways, depending on the context, for example:

- compressing mappings, that is $f : X \to Y$, such that for all $x, y \in X$ the inequality $d_Y(f(x), f(y)) \le d_X(x, y)$ holds. Isomorphisms in this category are isometries, that is, bijections preserving distances,
- Lipschitz mappings, that is such that $f : X \to Y$, that is there exists a constant $C = C(f)$, such that for all $x, y \in X$ the inequality $d_Y(f(x), f(y)) \le Cd_X(x, y)$ is true. Isomorphisms in this category are bilipschitz mappings, i.e., mappings for which
- $C_1d_X(x, y) \le d_Y(f(x), f(y)) \le C_2d_X(x, y)$ is true for every $x \in X$ and every $y \in X$.
- Hölder mappings, that is, $f : X \to Y$ for which there is a constant $C = C(f)$, and
  $a = a(f) \le 1$ such that for all $x, y \in X$ the inequality $d_Y(f(x), f(y)) \le Cd_X(x, y)^a$ is true.
- The product in the category of metric spaces is the metric space

$(X, d_X) \times (Y, d_Y) = (X \times Y, d_{X\times Y})$, where $d_{X\times Y}$ can be defined as follows, for example:

$$d_{X\times Y}((x_1, y_1), (x_2, y_2)) = \left(d_X(x_1, x_2)^C + d_Y(y_1, y_2)^C\right)^{\frac{1}{C}}.$$

For $C = \infty$, we get the Chebyshev metrics:

$$d_{X\times Y}((x_1, y_1), (x_2, y_2)) = \max\{d_X(x_1, x_2), d_Y(y_1, y_2)\}.$$

For $C = 1$, we get a taxi-cab (Manhattan) metrics:

$$d_{X\times Y}((x_1, y_1), (x_2, y_2)) = d_X(x_1, x_2) + d_Y(y_1, y_2).$$

For $C = 2$, Euclidean metrics is obtained:

$$d_{X\times Y}((x_1, y_1), (x_2, y_2)) = \sqrt{d_X(x_1, x_2)^2 + d_Y(y_1, y_2)^2}.$$

So, we have:

We can construct a map from objects of Group category of all small (finite) groups $Gr(f)p$ into Graph category of all small graphs **Grph** associating to every group G its Cayley graph $\Gamma(G, M)$, where $M = G\backslash\{e\}$.

Then we associate to every homomorphism $f : G \to G'$ of a (finite) group G into a (finite) group G' a map φ from Cayley graph $\Gamma(G, M)$, where $M = G\backslash\{e\}$, into a Cayley graph $\Gamma(G', M')$, where $M' = G'\backslash\{e\}$, such that $\varphi(v_g) = w_{g'}$, if v_g is a vertex of a graph $\Gamma(G, M)$ associated with an element $g \in G$, then $w_{g'}$ is a vertex of a graph $\Gamma(G', M')$, such that $f(v_g) = w_{g'}$.

So, we have a functor from of all small (finite) groups $Gr(f)p$ into a category of all small (complete finite) graphs **Grph(cf)**.

We can construct a map from objects of category of all small groups **Grp** into a set of objects of a category of all small (complete) graphs **Grph(cf)**.

We can construct a map from objects of category of all small graphs **Grph** into a set of objects of a category of all Euclidean spaces **Eucs**.

Here we have used the following designations:

Graph category of all small graphs **Grph**

Group category of all small groups **Grp**

Category of all Euclidean vector spaces **Eucs**

Here from we have that group category can be embedded into graph category by means of Cayley graphs of the group [6, see Joneda's lemma]. The category of metric spaces is embedded in the category of graphs.

Let's remind the following definition.

Definition 8.14 An object Q of a category $\mathfrak{C}$ is called an injective object if for every morphism $f : A \to Q$ and every monomorphism $h : A \hookrightarrow B$ (designation $\hookrightarrow$) there exists a morphism $g : B \to Q$, continuing f, that is $g°h = f$:

See the above figure.

8.3.2   The Relationship Between Groups and Graphs from the Outlook of Category Theory

Cayley group graph is a graph that is built on a group with a dedicated system of generators. Let's suppose that a group G is given and its system of generators S that is for every $s \in S$ it is true $s^{-1} \in S$. The Cayley graph of a group G by the system of generators S is a graph whose vertices are the elements of the group and the element

$g \in G$ gg is connected by an edge exactly to those elements that are obtained by multiplying g by an element of S.

In Chap. 7, we presented the construction of a multidimensional graph in a Euclidean vector space (Definition 7.1) and showed how the Cayley graph $\Gamma(G, M)$ of a group G, where M is the system of generators of a group G, can be transformed into the graph $\Gamma_{(G,M)}$ with labels on the edges in an n—dimensional vector space R^n

Namely, Chap. 7:

Let's consider Cayley graph $\Gamma(G, M)$ of a group G, where M is the system of generators of a group G.

Let's then consider an epimorphism φ of a free group $F_{\omega_1} = \langle F_{\omega_1}, *, ^{-1}, e \rangle$ of an uncountable rank onto an additive group $\langle R^n, +, -, 0 \rangle$ of an Euclidean vector space R^n: $\varphi : F_{\omega_1} \to R^n$.

The ends of the edges of the Van Kampen diagram $D(G)$ correspond to the vertices of the graph $\Gamma_{D(G)}$, thus connected with the Van Kampen diagram $D(G)$. Then each edge $\left(V^1_{\alpha_1}, V^1_{\beta_1} \right)$ of the graph $\Gamma_{D(G)}$ corresponds some label a_i, written on this edge of the Van Kampen diagram. So, one gets that in this way the Van Kampen diagram $D(G)$ is transformed into the graph $\Gamma_{D(G)}$ with labels on the edges in the n—dimensional vector space R^n. A graph $\Gamma_{(G,M)}$ will be a complete one if and only if $M = G \backslash \{e\}$ [10].

The graph $\Gamma_{(G,M)}$ (Chap. 7), will be an analogue of the injective object in the category of all graphs. The graph $\Gamma_{(G,M)}$ for different groups G, constructed by the Van Kampen lemma is complete for any group G, provided that the system of generators M of the group G satisfies the condition: $M = G \backslash \{e\}$, but with different the number of vertices, and different marks on the edges. This is a possible explanation for the fact that with the help of Van Kampen diagrams practically all problems of group theory are solved.

8.3.3 Van Kampen Probability Diagram (Cayley Group Graph). Probability-Isomorphic Groups

We need here to recollect Erdés–Renyi algorithm and the description the Erdés–Renyi random graphs model, see Chap. 7. Following this model, let's embed the Van Kampen diagram $D(G)$ (Cayley graph $\Gamma D(G)$) of a finite, finite quasi-fractal group G transformed into a graph with labels on edges into an n—dimensional vector space R^n. After that let's apply the Erdés–Renyi algorithm and obtain from this the definition of probability-isomorphic groups by binomial distribution.

Definition 8.15 Finite groups (quasi-fractal finite groups) $G = \langle G, *, ^{-1}, e \rangle$ and $G' = \langle G', \circ, ^{-1}, e' \rangle$ are called probability-isomorphic ones by binomial distribution if by the model of Erdés–Renyi the probability of the fact that Cayley graph $\Gamma D(G)$ with $M \subset G, M = G \backslash \{e\}$, of a group G and Cayley graph $\Gamma D(G')$ with $M' \subset G', M' = G' \backslash \{e\}$, of a group G' are equal (isomorphic) is more than 0.5.

Theorem 8.16 [3] Any two finite groups of the same order are probability-isomorphic by regard to the binomial distribution.

Proof Cayley graph $\Gamma D(G)$ with $\subset G, M = G\backslash\{e\}$, of a group G and Cayley graph $\Gamma D(G')$ with $M' \subset G', M' = G'\backslash\{e\}$, of a group G' where G and G' are finite groups of the same order are isomorphic to each other.

Theorem 8.17 Any two fractal finite groups of the same order are probability-isomorphic by regard to the binomial distribution.

Proof The theorem is a direct consequence of Theorem 8.16.

One can see from here the analogy between probability-isomorphic finite quasi-fractal groups and giant components.

References

1. Home page. https://en.wikipedia.org/wiki/Giant_component
2. Mac Lane, S.: Categories for the Working Mathematicians, 2nd edn. Springer, New York (1998)
3. Palyutin, E.A.: Interpretation of graphs in noncommutative theories of Frechet degrees. Fund. Appl. Mathem. **15**(2), 145–167 (2006)
4. Jäkel, C.: A Unified Categorical Approach to Graphs. Technische Universität Dresden (2015)
5. Matumoto, T.: Any group is represented by an outer automorphism group. Hiroshima Math. J. **19**, 209–219 (1989)
6. Prze'zdziecki, A.J.: An "almost" full embedding of the category of graphs into the category of groups. Adv. Mathem. **225**(4), 1893–1913 (2010)
7. Grossman, I., Magnus, W.: Groups and Their Graphs. Random House, The L.W. Singer Company (1964)
8. Olshansky, AYu.: The geometry of defining relations in groups. Nauka Publishers, Moscow (1989). (in Russian)
9. Lindon, R.C., Schupp, P.E.: Combinatorial Group Theory. Springer-Verlag, Berlin, Heidelberg, New York (1977)
10. Home page, ru.bstu.wiki Vertex-Transitive Graph - National Library named after R.E. Bauman
11. Home page. File: Cayley graph of F2.svg
12. Home page. https://en.wikipedia.org/wiki/Van_Kampen_diagram
13. Whitney, H.: Congruent graphs and connectivity of graphs. Amer. J. Math. **54**, 150–168 (1932)
14. Harari, F.: Graph theory, Mir, Moscow (1973) (in Russian)
15. Serdyukova, N.A.: On generalizations of purities. Algebra Logic **30**(4), 432–456 (1991)
16. Zolotarev, V.M.: Probabilistic metrics. Prob. Theory Appl. **18**(2), 264–287 (1983)
17. Cobham, A.: Undecidability in group theory. Notices Am. Math. Soc. **9**, 406 (1962)

---

[3]The complexity of the question. It turned out [Cobham, 17] that the theory of finite groups is not even recursively axiomatizable.

Chapter 9
Smart System Sustainability and Smart System Management

> However, in mathematics there are no coincidences.
> Mathematics is the realm of necessity.
> Even the most superficial and casual in appearance coincidence
> always expresses a deeper analogy.
> N. Vavilov, Concrete group theory.
> http://mathscinet.ru/files/VavilovN.pdf

Abstract Recently the concepts of smart systems, smart education, smart engineering, smart medicine, smart city, smart universities, smart technology and smart communication were introduced in scientific methodology and practice for a modern society. All of these entities are the components of a new system—smart society system. These components must be equipped with corresponding connections that ensure their reliability, sustainability and security, or, in other words, to have an appropriate smart infrastructure. It turns out that we approach the concept of smart infrastructure from the standpoint of tensor estimation. So, the main questions to consider here are the following ones:

- verbal tensor estimates in the Smart Systems Theory,
- methods to increase system reliability,
- tensor estimation of system Q—quality,
- the algorithm of tensor estimation of system Q—quality,
- formalization the approaches and the concept of system sustainability,
- the algorithm of tensor estimation of system sustainability,
- sustainability of a quasi-fractal system (a system modeled by a quasi-fractal algebraic system),
- system sustainability violation: violation of the system's closeness property; malfunction of one of the system's factors,
- violation of feedback in the system; factor-flexible quasi-fractal system,
- substitution of system's functions; system's compensational possibilities,
- compensational functions of a factor-flexible quasi-fractal system.

Keywords System sustainability · Quasi-fractal system · System management

© Springer Nature Switzerland AG 2021

N. A. Serdyukova and V. I. Serdyukov, *Algebraic Identification of Smart Systems*, Intelligent Systems Reference Library 191,
https://doi.org/10.1007/978-3-030-54470-6_9

9.1 Introduction. Verbal Tensor Estimates in the Smart Systems Theory

In [1] and also in the works [2, 3] the issues of sustainability, compensatory stability, stability of smart systems have been considered. In this chapter, we shall consider sustainability issues in relation to systems whose models are quasi-fractal algebraic systems. At the same time, the sustainability of such systems will be considered in line with the results of (Chap. 3, Sect. 3.4.1 Geometric Examples) using the technique of example 4 and the concept of regulating quasi-fractal system.

Let's begin with examples.

Example 1 Consider the very phrase SMART,[1] (**Specific, Measurable, Attainable, Relevant, Time-bound**), [4]. This phrase or abbreviation in fact, is about the need to use for tensor estimation in management and project management. Moreover, the phrase SMART, in fact, is a verbal model of the system, written in the short form.

Example 2 Let's consider SWOT analysis, [5]. SWOT analysis can be used to build organizational or personal strategy, for Matching and converting, corporate planning, marketing. According to [5], SWOT (**Strengths**: characteristics of the business or project that give it an advantage over others).

Weaknesses: characteristics of the business that place the business or project at a disadvantage relative to others. **Opportunities**: elements in the environment that the business or project could exploit to its advantage. **Threats**: elements in the environment that could cause trouble for the business or project analysis (or SWOT matrix) is a strategic planning technique used to help a person or organization identify strengths, weaknesses, opportunities, and threats related to business competition or project planning. This phrase or abbreviation, too, in fact, is about the need to use for tensor estimation in management and project management.

In [1, Chap. 6] we have constructed a tensor estimate of the efficiency of the functioning of smart system S modelled by a group of factors G_S. This was necessary because smart systems are complex and there are usually several criteria for their effectiveness. Thus, it was a question of determining the effectiveness of a multicriteria model. Moreover, the phrase SWOT, in fact, is a verbal model of the analysis of a system, written in the short form.

Using of SWOT Analysis. According to [6], SWOT Analysis which has been used over the last fifty years in the field of strategic management is a valuable technique for planning and decision making. There we can read the following. In the 1980's the SWOT analysis was used in the Small and Medium Enterprises development and the creation of business and marketing plans. Weihrich (1982) introduced TOWS Matrix for matching the environmental threats and opportunities with the organization's weaknesses and especially its strengths. He matched the SWOT variables in a systematic fashion. In the 1990's the SWOT method was also used in regional

---

[1]The first known use of the term is found in Paul J. Meyer in 1965 and later in November 1981 in Management Review by Doran [4].

development, project formulation and the social marketing of Non-Governmental Organizations. Dealtry (1992), developed 'Dynamic SWOT Analysis' namely DSA. He designed SWOT as a dynamic project management process involving managers in the preparation and implementation of value driven strategies using learning organization principles, ideas and cycles. It starts with raw shock analysis and moves onwards through group and individual activities such as reading the dashboard of development, determining the metrics of pulse and threshold issues, decision analysis and prioritizing SWOT attributes as a basis for taking executive action and tactical project management. Wheelen and Hunger (1998) developed SFAS Matrix-Strategic Factors Analysis Summary-plus EFAS-External Factor Analysis Summary-Table and IFAS-Internal Factor Analysis Summary-Table to deal with the criticism of SWOT. SFAS Matrix, summarizes strategic factors by combining the external factors from the EFAS Table with the internal factors from the IFAS Table. Possible alternative strategies can then be generated by referring to EFAS and IFAS tables in generating SWOT Matrix. After the 1990's, the criticisms directed to the concept of strategic planning were also directed to SWOT Analysis. These criticisms contributed to the development of SWOT Analysis. A number of scholars have proposed variants of SWOT to enrich the planning process. One such variant is WOTSUP, where UP means 'Underlying Planning' and the other variant is SOFT, where weaknesses have been re-identified as 'Faults'. Many scholars suggest the need to use additional analysis instead of SWOT or using it in combination with other techniques. Complimentary analysis techniques to formulate SWOT based strategies can be regarded as Kaplan and Norton's Balanced Score Card, Quality Function Deployment-QFD methodology with Balanced Score Card, Cross Impact Analysis, Porter's Five Forces Model, Porter's Generic Strategies, scenario analysis, Mc Kinsey's 7S Framework, AHP, ANP, Fuzzy AHP-FAHP-, Fuzzy ANP, Axiomatic Fuzzy Set-AFS-Theory with Evidential Reasoning-ER-, Multiple Criteria Group Decision Making—MCGDM-method with nonhomogeneous preference information, SWOT with SMART-simple multi-attribute rating technique-method, SWOT with SMAA-O-the stochastic multi-criteria acceptability analysis with ordinal criteria-method, SWOT with MADM-multi-attribute decision making-technique based on the concept of grand strategy matrix, PEAK-power, earning, artistry or activity and knowledge with SETS-solidity, extent, type and segment or specialization, Radar Qualitative Screen, Radar Quantitative Screen, TRIZ-Theory of Inventive Problem Solving, Delphi Panel with descriptive statistical analysis, SWOT and CSM-contrast mining application. Alternatives to a conventional SWOT Analysis include VRIO-valuable, rare, imperfectly imitable, organizational oriented-Framework, Goals Grid, Telescopic Observations Strategic Framework, SCORE-strengths, challenges, options and opportunities, responses, effectiveness-Analysis, SORF-strengths, opportunities, realities, facts-Analysis.

Example 3 Sustainability of complex organizations

Now let's consider sustainability of complex organizations.
Let's start with the definitions of the basic concepts.
In [3] the following remarks were done:

For organizations of a complex structure operating in a competitive environment, stability should be understood as the complex property of the system [control object (CO) and its control system (CS)], characterized by the following indicators: **surviv-ability**—the ability of the system to perform tasks under the deliberate effect of all means of destruction from competitors; **stability**—the ability of the system to perform tasks in emergency situations; **reliability**—the ability of the system to perform tasks, maintaining the operability and quality of functioning for a given time (during operation). So, we have abbreviation SSR. This phrase or abbreviation in fact, is about the need to use for tensor estimation for the concept of sustainability. Moreover, the phrase SSR, in fact, is a verbal model of the sustainable system, written in the short form. Here the author tries to highlight the main features of reliability by introducing the term "survivability" of a complex informational—technological system.

9.1.1 Methods to Increase System Reliability

In [2] one of the ways to increase the reliability of products is considered due to mixed redundancy with recovery in combination with the use of elements of artificial intelligence. Products reliability function is determined and investigated, advantages and disadvantages are evaluated, and the advantage of separate redundancy is proved, that implies the presence of a switch that operates in automatic mode. The article presents a system of equations describing the development of a random process, and its solution, followed by analysis of the results, which demonstrate the advantages of mixed redundancy with recovery. The mathematical apparatus proposed allows one to analyze the duration of uptime of products and the expedience to complicate the design of systems.

In [7] another view on system reliability is presented that is: the system reliability quality is in deep connection with an appropriate system smart infrastructure.

One of the most important issues in this regard is a feedback and a control in smart system. So, in [7] the common approach to an application of statistical methods to automated knowledge control is described. The obtained research outcomes clearly show that if testing is used to test knowledge level, then (a) it gives adequate outcomes of knowledge assessment only under repeated regular use of testing procedure and (b) the final assessment should be obtained using the averaging procedure.

All these methods use a large number of numerical indicators related to the esti-mated quality of the system. The problem arises of constructing a tensor quality estimate of the Q—property functioning of the system, or functioning of the system, according to Q—property, where Q is some quality of a system S.

9.2 Tensor Estimation of System Q—Quality

Sometimes one needs to have a complex estimation of the system concerning some qualitative sides of its functioning. For example, such situation one has when considering the efficiency of system functioning, the reliability of system functioning, the sustainability of system functioning.

Let Q be some quality of a system S, which can be estimated with a help of a set

$$H_Q(S) = \{h_1, h_2, \ldots, h_n\}$$

of numerical characteristics. We shall assume, that these numerical characteristics are functions depending of the time: $h_i = h_i(t), t = 1, \ldots, n$.

We shall construct a tensor estimate of the Q—quality of the functioning of the system S as a homomorphism of a group of factors G_S, determining the system S into a group $GL(n, R)$ of linear homogeneous transformations of the vector space R^n. We shall use the same construction as in [1].

In this connection we recall how a complete linear group is defined. We get the possibility of constructing various tensor estimates of the system's functioning Q—quality by considering the homomorphisms of the group G_S, that define a system S into a group $GL(n, R)$ of linear homogeneous transformations of the vector space Δ^n, where Δ is a finite field or a field different from R.

Let's recall the following definition.

Definition [8] A group $GL(n, \Delta)$ is called a full linear group of degree n over the ring Δ. It is isomorphic to the multiplicative group of the ring Δ_n, where Δ_n is a ring of all $n \times n$—matrices over the ring Δ.

Then $\Delta_n \cong End\,M$ where M is a free right Δ—module with the finite basis $u_1, \ldots, u_n$. If $\sigma \in End\,M$, then $\sigma(u_\nu) = \sum\limits_{\nu=1}^{n} u_\nu \, \alpha_{\mu\nu}, \alpha_{\mu\nu} \in \Delta, \nu = 1, \ldots, n$. Let $A = [\alpha_{\mu\nu}]$. A matrix A is called a matrix of endomorphism σ in the basis $u_1, \ldots, u_n$. Then the map $\varphi: End\,M \to \Delta_n$ such that $\varphi: \sigma \mapsto A$, is an isomorphism.

From [1] we have:

To evaluate the sustainability of a system Q—quality we construct a tensor estimate. We represent the tensor estimate as a homomorphism

$G_S \to GL(n, R)$, where G_S is a group of factors defining the system, $GL(n, R)$ is a group of linear homogeneous transformations of a vector space R^n, n—a number of quantitative indicators that assess the system Q—quality of the system S.

Let's assume that G_S is a finite group, consisting of n elements: $G_S = \{a_1, a_2, \ldots, a_n\}$ and $h_i = h_i(t) = h_i(a_i, t)$ is a numerical characteristic of a factor a_i.

So, a homomorphism $f: G_S \to GL(n, R)$ can be considered as the tensor estimation of the quality Q.

Definition 9.1 A homomorphism $f: G_S \to GL(n, R)$ can be considered as the tensor estimation of the quality Q.

Another question which arises in this connection is the question about the properties of this estimation.

9.2.1   The Algorithm of Tensor Estimation of System Q—Quality

Algorithm 9.1

(1)   The construction of quantitative indicators describing the functioning of the system with regard to Q—quality. Identifying and monitoring the compliance of the functioning of the system with its purpose.

(2)   The construction of the tensor evaluation of the system's functioning with regard to Q—quality.

(3)   The identification and monitoring of all the links of the innovation system with the external supersystem and its subsystems that arise during the functioning of the system with regard to Q—quality.

(4)   The identification of internal and external attributive factors of the system with regard to Q—quality.

(5)   The construction of internal qualitative indicators of the functioning of the innovation system in the form of a graph and a group of internal attributive features of the innovation system.

(6)   The construction of external quality indicators of intrasystem connections of the system in the form of a graph and a group of external attributive features of the system with regard to Q—quality.

(7)   The determination of the number of possible synergetic effects of the system with regard to Q—quality.

(8)   The construction a model a system's functioning with regard to Q—quality based on its internal and external attributive features.

(9)   The verification of the possibility of increasing the accuracy of the model by introducing additional factors into the model.

(10) The correction the management actions to prevent the occurrence of undesirable synergistic effects.

Now we shall try to investigate some concrete system properties from the point of view of their assessment. One of the main system properties is a property of system sustainability. Let's start with the description of the approaches and the concept of the system sustainability.

9.3 Formalization the Approaches and the Concept of System Sustainability

The concept of sustainability is well—studied in terms of the availability of various quantitative parameters describing the dynamic behavior of the system, [1, 2]. There were introduced such concepts as Lyapunov sustainability, Zhukovsky sustainability and so on. We shall consider discrete systems.

Under the sustainability of a discrete system we shall understand its ability to return to the equilibrium position after the end of the action of external factors as in the case of continuous—time systems.

In [1] it was shown that the notion of a final sustainable system is in fact an analogue of the classical notion of a sustainable equilibrium.

Just as in [1, 9], we understand system stability to mean the following:

1. We consider scalar discrete dynamical system:

   This discrete dynamical system S is defined by the map f:

$$u \mapsto f(u) = f(u,r), u \in R, r \epsilon R^m, f: R \to R, \tag{9.1}$$

that is:

2. Let $N_t, t \in T$, be an indicator describing the system S.
3. Suppose that there is a ratio.

$$N_{t+1} = f(N_t), N_t \in R, f: R \to R \tag{9.2}$$

or equivalently,

$$N \to f(N), N \in R, \tag{9.3}$$

where f is a given map, depending on a number of parameters.

Equations (9.2) and (9.3) defined the scalar discrete dynamical systems S.

4. A fixed point u^* of the map (9.1) is called a sustainable one if for every $\varepsilon > 0$ there exists such $\delta > 0$ that for any initial data u_0 from the δ—neighborhood of the point u^* the whole system's trajectory $u_t, t = 0, 1, 2, 3, \ldots$ is contained in the ε—neighborhood of the point u^*.
5. Under the sustainability of a discrete system S we shall understand its ability to return to the equilibrium position after the end of the action of external factors as in the case of continuous—time systems,

   Equations (9.2) and (9.3) defined the scalar discrete dynamical system.
   The following definitions are contained in [9].

Definition 9.2 The phase space or a state space of a system S is the set of all its states $\{N_t | t \in R\}$, (9.2).

Definition 9.3 A trajectory or an orbit of a system, (9.1), generated by the map f is a sequence of points $\{N_t | t = 1, 2, \ldots\}$.

Definition 9.4 The points N^* from the set of states $\{N_t | t \in R\}$, (9.2), of a system S, such that $f(N^*) = N^*$ are called fixed points of the system S.

9.3.1 The Algorithm of Tensor Estimation of System Sustainability

Here we shall repeat algorithm of tensor estimation of system Q—quality for the case where Q is a property of system sustainability. Let's enumerate the indices of a system sustainability at first.

We need the following definition the notion of a final sustainable system, which in fact is an analogue of the classical notion of a sustainable equilibrium from [1].

Definition 9.5 [1] Let $G_S = \langle G_S, \circ, \ ^{-1}, e \rangle$ be a group of factors which represent the system S. Let G_S be finite and $|G_S| = n$. Let $G_{1S}, G_{2S}, \ldots, G_{mS}$ be all pair wise non isomorphic groups of n elements. Groups $G_{1S}, G_{2S}, \ldots, G_{mS}$ are called final states of a system S. A system S is called a final sustainable one if it has only one final state.

Definition 9.6 [1] Groups $G_{1S}, G_{2S}, \ldots, G_{mS}$ are also called scenarios of a development of a system S.

The following statements follow directly from the Definition 9.5.

Corollary 9.7 [1] System S, which is represented by a simple group of factors G_S, where the number of factors in G_S, is equal or less then 10^6, is final sustainable.

The loss of system sustainability can be connected with the following:

- a system has more than one final state,
- a system loses its closed nature.

The possible indicators for tensor estimation of a system sustainability are the following ones: sustainability static indices and sustainability dynamic indices.

So, sustainability static indices are:

- the number of final states of a system,
- the indices of loses the system closeness.

Sustainability Dynamic Indices

Let's move on to the sustainability indices of dynamic systems. Before doing this, note that in [10] with respect to linear dynamical systems, it is also noted that the sustainability of a linear system (i.e., the system is described by a linear differential equation) is determined not by the nature of the perturbation, but by the system structure. The foundations of the strict sustainability theory of dynamical systems were developed by A. M. Lyapunov in "the General problem of the sustainability of motion" (1892). The transfer function of any linear dynamic system can be represented as the ratio of polynomials from which the operator equation of the system can be obtained. Then from the operator equation of the system one can go to the differential equation of the system and to the characteristic equation corresponding to this differential equation.

For automatic control systems (ACS), for example, algebraic sustainability criteria of linear dynamic systems are used. The signs of the real parts of the roots of the characteristic equation corresponding to the differential equation of the system are the sustainability indicators of linear dynamic systems in this case.

The sustainability condition of the linear dynamic ACS is formulated as follows:

For a system to be sustainable, it is necessary and sufficient that all the roots of its characteristic equation will have a negative real part.

So, sustainability dynamic indices are:

- the characteristic function of coincidences the number of roots of characteristic equation corresponding to the differential equation of the system to the number of roots having a negative real part.

Algorithm 9.2

(1) The construction of quantitative indicators describing the functioning of the system with regard to sustainability. Identifying and monitoring the compliance of the functioning of the system with its purpose.
(2) The construction of the tensor evaluation of the system's functioning with regard to sustainability.
(3) The identification and monitoring of all the links of the innovation system with the external supersystem and its subsystems that arise during the functioning of the system with regard to sustainability.
(4) The identification of internal and external attributive factors of the system with regard to sustainability.
(5) The construction of internal qualitative indicators of the functioning of the innovation system in the form of a graph and a group of internal attributive features of the innovation system.
(6) The construction of external quality indicators of intrasystem connections of the system in the form of a graph and a group of external attributive features of the system with regard to sustainability.
(7) The determination of the number of possible synergetic effects of the system with regard to sustainability.

(8)   The construction a model a system's functioning with regard to sustainability based on its internal and external attributive features.
(9)   The verification of the possibility of increasing the accuracy of the model by introducing additional factors into the model.
(10) The correction the management actions to prevent the occurrence of undesirable synergistic effects.

Now let's consider the violation of the sustainability property of the system.

9.4 Sustainability of a Quasi-fractal System (A System Modeled by a Quasi-fractal Algebraic System)

In this section we spread the notion of sustainability to the quasi-fractal systems. We shall consider quasi-fractal systems modeled by quasi-fractal groups.

Once again, we are under conditions of the Definition 2.1 from Chap. 2.

Definition 2.1 [Chap. 2]

Let's consider the algebraic system $A_1 = \langle A_1; \Omega_1 \rangle$ of the signature Ω_1, such that every element a_α, $\alpha \in \Lambda_1$, of the main set A_1 of the system A_1 in turn is an algebraic system of the signature Ω_2. That is $a_\alpha = A_2^\alpha = \langle A_2^\alpha; \Omega_2 \rangle$, $\alpha \in \Lambda_1$ is an algebraic system of the second level. Continue this process by induction. If an algebraic system $A_k^\alpha = \langle A_k^\alpha; \Omega_k \rangle$, $\alpha \in \Lambda_k$ is an algebraic system of the level k of the fractal and every element a_α, $\alpha \in \Lambda_k$ of the main set A_k^α of the system A_k^α is an algebraic system $a_\alpha = A_{k+1}^\alpha = \langle A_{k+1}^\alpha; \Omega_{k+1} \rangle$ of the signature Ω_{k+1}, of the level $k+1$, $\alpha \in \Lambda_{k+1}$, of the fractal, then the algebraic system $A_1 = \langle A_1; \Omega_1 \rangle$ is called a quasi-fractal algebraic system. If all signatures Ω_k, $k = 1, \ldots n, \ldots$, are equal to each other and all the systems $\langle A_k^\alpha; \Omega_k \rangle$, $\alpha \in \Lambda_k$, are isomorphic to each other then the algebraic system $A_1 = \langle A_1; \Omega_1 \rangle$ is called a fractal algebraic of the signature Ω_1.

We should explain that in this notation $A_1 = \langle A_1; \Omega_1 \rangle$ plays the role of a universal variable to denote a quasi-fractal algebraic system.

Let's note that an ordinary algebraic system is a quasi-fractal system of the first level.

Now let's assume that quasi-fractal system $G_S = \langle G_S; *, {}^{-1}, e \rangle = A_1 = \langle A_1; \Omega_1 \rangle$ is a quasi-fractal group, that is at each level $A_k^\alpha = \langle A_k^\alpha; \Omega_k \rangle$, $\alpha \in \Lambda_k$ of a quasi-fractal system $A_1 = \langle A_1; \Omega_1 \rangle$ one has a group: $G_k^\alpha = \langle G_k^\alpha; \Omega_k \rangle = A_k^\alpha = \langle A_k^\alpha; \Omega_k \rangle$.

Definition 9.8 Let $G_S = \langle G_S, \circ, {}^{-1}, e \rangle = A_1 = \langle A_1; \Omega_1 \rangle$ be a quasi-fractal group of factors which represent the system S. Let $G_k^\alpha = \langle G_k^\alpha; \Omega_k \rangle$ be finite for every quasi-fractal level k and $\left| G_k^\alpha \right| = n_k$. Let $G_{1S}(k), G_{2S}(k), \ldots, G_{mS}(k)$ be all pair wise non isomorphic groups of n_k elements. Sequence of groups

$< (G_{1S}(k), G_{2S}(k), \ldots, G_{mS}(k))|k = 1, \ldots, n, \ldots >$ is called a final state of a system S according the model $G_S = \langle G_S, \circ, \ ^{-1}, e \rangle = A_1 = \langle A_1; \Omega_1 \rangle$. A system S is called a final sustainable one according the model $G_S = \langle G_S, \circ, \ ^{-1}, e \rangle = A_1 = \langle A_1; \Omega_1 \rangle$ if it has only one final state.

9.5 System Sustainability Violation: Violation of the System's Closeness Property. Malfunction of One of the System's Factors

In this section we shall follow [1].

From [Theorem 3.1, 1, Chap. 3] we have:

Theorem 9.9 Upon termination of the functioning of one of the factors from the group $G_k^\alpha = \langle G_k^\alpha; \Omega_k \rangle$ of order $|G_k^\alpha| = n_k^\alpha$ representing the level k of the quasi-fractal system S that is equivalent to the removal of this factor from the group of factors G_k^α, the system S loses the property of closeness on the level k.

The proof is obtained directly from Lagrange theorem, as the number of factors continue to operate in the system S, is $n_k^\alpha - 1$ and $n_k^\alpha - 1$ does not divide n_k^α.

Theorem 9.10 Upon termination of the functioning of m factors where m does not divide n_k^α, from the group G_k^α of order $|G_k^\alpha| = n_k^\alpha$ representing level k of the quasi-fractal system S that is equivalent to the removal of these factors from the group of factors G_k^α, the quasi-fractal system S loses the property of closeness on the level k.

Proof In this case $n_k^\alpha - m$ does not divide n_k^α, and according to Lagrange theorem there is no a subgroup of order $n_k^\alpha - m$ in G_k^α, so the system S loses the property of closeness on the level k.

Theorem 9.11 A quasi-fractal system S with the group of representing factors G_k^α of the level k of the system S retains the closure property on the level k after the cessation of the functioning factors $\{a_i | i \in I_k^\alpha\}$ if and only if $\langle G_k^\alpha \backslash \{a_i | i \in I_k^\alpha\}, \circ, \ ^{-1} \rangle$ is a group.

In [1, Chap. 3] there are the simplest examples to illustrate these statements.

In the same way, the theorems from [1, Chap. 3, Sect. 3.2, pp. 44–47] can be reformulated for levels of a quasi-fractal system.

So, let's consider some examples of violation of feedback in the quasi-fractal system.

9.6   Violation of Feedback in the System. Factor-Flexible Quasi-fractal System

We begin with the following theorem, following from [1, Chap. 3, Sect. 3.2, pp. 44–47].

Theorem 9.12 Let the system S be modeled by a quasi-fractal group of permutations of the third degree $S_3 = \langle S_3, *, \ ^{-1} \rangle = A_1 = \langle A_1; \mathit{\Omega}_1 \rangle$. At any level of the quasi-fractal representing this system it is not possible to regulate exactly:

- one factor representing the system,
- two factors representing the system,
- four factors representing the system.

When regulating precisely:

- one factor representing the system,
- two factors representing the system,
- four factors representing the system,

system S will change its structure.

Theorem 9.13 Upon termination of the functioning of one of the factors a or a^{-1} from the group G_k^α representing level k of the quasi-fractal system S, the system S loses the property of feedback on this factor on the level k.

This can be interpreted in another way: the signature of the system on level k changes that is the group signature $\langle *, \ ^{-1}, e \rangle$ changes for semigroup with unit signature $\langle *, e \rangle$, and one get semigroup $\mathbf{Smg}_{k,\alpha}$ which represents the level k of the quasi-fractal system S instead of a group of factors G_k^α, which represents the level k of the quasi-fractal system S.

Theorem 9.14 Let closed associative system S with a feedback be simulated by a quasi-fractal group of factors isomorphic to alternating group $A_n, n \geq 5$. Upon termination of any of the factors determining the system S, with probability equals to p_1, any subsystem of a system S changes its functioning with probability equals to p_1.

Proof The alternating group $A_n, n \geq 5$, is a simple group, that is it does not have normal subgroups other than the entire group and the unit subgroup.

In the same way, the definitions and theorems from [1, Chap. 3, pp. 49–54] can be reformulated for levels of a quasi-fractal system.

9.6.1 Substitution of System's Functions. System's Compensational Possibilities

By replacement (compensation) of (disturbed) functions of the system, we shall mean the adaptation of the system to the changed conditions for the existence and replacement of the nonworking (failed, ineffectively working) elements of the system with respect to the elements of the system that are operating (possibly more efficiently than failed system's elements). Such elements of the system will be called substitutional or compensatory.

Now we are again under the conditions of the Definition 9.8. So, under the conditions of the Definition 9.8 we have:

Let $G_S = \langle G_S, \circ, \,^{-1}, e \rangle = A_1 = \langle A_1; \Omega_1 \rangle$ be a quasi-fractal group of factors which represent the system S. Let $G_k^\alpha = \langle G_k^\alpha; \Omega_k \rangle$ be finite for every quasi-fractal level k and $|G_k^\alpha| = n_k$. Let $G_{1S}(k), G_{2S}(k), \ldots, G_{mS}(k)$ be all pair wise non isomorphic groups of n_k elements. Sequence of groups $< (G_{1S}(k), G_{2S}(k), \ldots, G_{mS}(k))|k = 1, \ldots, n, \ldots >$ is called a final state of a system S according the model $G_S = \langle G_S, \circ, \,^{-1}, e \rangle = A_1 = \langle A_1; \Omega_1 \rangle$. A system S is called a final sustainable one according the model $G_S = \langle G_S, \circ, \,^{-1}, e \rangle = A_1 = \langle A_1; \Omega_1 \rangle$ if it has only one final state.

Let S_k the quasi-fractal level k of the system S be consisted of the following elements: $S_k = \{s_\alpha^k | \alpha \in \Lambda_k\}$. Let's suppose that the main set of the group of factors determine the system S_k is $G_{Sk} = \{e, a_{1k}, \ldots, a_{nk}\}$, in this case, the elements of the system S_k, that determine each factor, i.e. there is a one-to-one correspondence are outlined: $a_{ik} \leftrightarrow S_{ik} = \{s_{\alpha_i}^k | \alpha_i \in \Lambda_i, i = 1, \ldots, n_k\} \neq \emptyset$. Let's note to every subset $S_{ik} = \{s_{\alpha_i}^k | \alpha_i \in \Lambda_i, i = 1, \ldots, n_k\}$ of the set S_{ik} one to one corresponds the subset $\{\Lambda_i | i = 1, \ldots, n_k\}$ of the set Λ.

Let

$F_{Sk} = \{f_{ik} | i \in I_k\}$ be functions corresponding to the set of factors $\{a_{ik} | i \in I_k\}$ that determine the system S_k. As it was noted above we can assume without loss of generality that $\{a_{ik} | i \in I_k\}$ coincides with $G_{Sk} = \{e, a_{1k}, \ldots, a_{nk}\}$.

The following diagram helps to understand the mechanism of compensatory possibilities of the system (Fig. 9.1).

The diagram (9.1) can be explained in the following way if to adhere to the classical scheme of studying systems: system—elements—structure. One can consider the factor $a_{ik}, i \in I_k$ as a cloud of elements $S_{ik} = \{s_{\alpha_i}^k | \alpha_i \in \Lambda_i, i = 1, \ldots, n_k\}$, and after that to consider the structure of the factors $a_{ik}, i \in I_k$, or, that is the same the structure of clouds—clusters $S_{ik} = \{s_{\alpha_i}^k | \alpha_i \in \Lambda_i, i = 1, \ldots, n_k\}$.

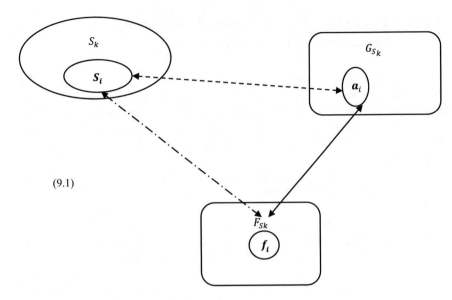

(9.1)

Fig. 9.1 Mechanism of compensatory possibilities of the system

9.6.2 Compensational Functions of a Factor-Flexible Quasi-fractal System

Let's consider the set of functions $F_{S_{ik}} = \{f_{ik}|i \in I_k\}$ of the quasi-fractal system S, which one to one correspond to the set of factors $G_{Sk} = \{e, a_{1k}, \ldots, a_{nk}\}$, that determine the level k of the quasi-fractal system S, which consists of elements $\{s^k_{\alpha_i}|\propto_i \in \Lambda_i, i = 1, \ldots, n_k\}$, that is $S_{ik} = \{s^k_{\alpha_i}|\propto_i \in \Lambda_i, i = 1, \ldots, n_k\}$.

The system S_{ik} under the action of the function f_{ik} onto the system S_{ik} transfers into the state $f_{ik}(S_{ik})$ herewith the model of the system S_{ik} transformed into the model $f_{ik}(G_{Sk})$. During this transition, the following can occur:

(1) the system S_{ik} loses the associative property in the case if the action of the function f_{ik} neutralizes the factor a_{jk}, or some several factors together with the factor a_{jk},
(2) the system S_{ik} loses the associative property,
(3) the system S_{ik} loses the closeness property,
(4) in the system S_{ik} under the impact of the function f_{ik} the influence of the factor a_{jk} will increase and may be the influence of some several factors determine the system S_{ik}.

We shall consider a special case when after each action of the function $f_{ik}, i \in I_k$, the model $f_{ik}(G_{Sk})$ will be a group. We shall also assume that all the effects of the function f_{ik} onto the system S_{ik} in the model G_{Sk} can be written in the language NPC (Narrow Predicate Calculus) of the signature $\Omega_{S_{ik}} = \langle \cdot, {}^{-1}, e, \{a_{jk}|j \in I_k\}\rangle$. The set of actions of the function f_{ik} onto the system S_{ik} in the model G_{Sk} will be denoted by $Th(f_{ik})$ and will be called the elementary theory of the function f_{ik}, that is $Th(f_{ik}) = \{Q_1 \ldots Q_n\varphi(x_1, \ldots, x_n) | G^i_{S_{ik}} \models Q_1 \ldots Q_n\varphi(x_1, \ldots, x_n)\}$.

A formula $Q_1 \ldots Q_n\varphi(x_1, \ldots, x_n)$ is called a symmetric one if for any substitution $\tau \in S_n$, where S_n is a symmetric group of substitutions of the degree n we have the equivalence $\varphi(x_1, \ldots, x_n) \equiv \varphi(x_{\tau(1)}, \ldots, x_{\tau(n)})$. The set of all symmetric formulas of $Th(f_{ik})$ will be denoted by $SymmTh(f_{ik})$. The set of all formulas $Q_1 \ldots Q_n\varphi(x_1, \ldots, x_n)$ from $Th(f_{ik})$, such that for transposition $\tau = (i, J)$ the equality имеет $\varphi(x_1, \ldots, x_n) \equiv \varphi(x_{\tau(1)}, \ldots, x_{\tau(n)})$ is true will be denoted by $\tau Th(f_{ik})$, and will be called formulas symmetric by the transposition τ for the function $f_{ik}, i \in I_k$.

Definition 9.15 Let $\tau = (i, J)$ be a transposition of a degree n. We shall say that the factor a_{ik} replaces the factor a_{jk}, if for the transposition $\tau = (i, j)$ the equality $\varphi(x_1, \ldots, x_n) \equiv \varphi(x_{\tau(1)}, \ldots, x_{\tau(n)})$ is true where $Q_1 \ldots Q_n\varphi(x_1, \ldots, x_n) \in Th(f_{ik})$.

Definition 9.16 Let's say that the system S_{ik} admits the replacements of functions if there exists such transposition $\tau = (i, j)$ of a degree n_k, that $\tau Th(f_{ik}) \neq \emptyset$.

Hence from we obtain the following theorem.

Theorem 9.17 If for the group of factors $G_{S_{ik}}$ that determine the system S_{ik}, the inequality $\tau Th(f_{ik}) \neq \emptyset$ takes place that is if the group $G_{S_{ik}}$ has a symmetric formula representation then the system S admits the replacement of functions.

Let's connect the substitution of functions with the sustainability of the system.

Definition 9.18 Let us say that the system S_{ik} is sustainable by the factors a_{ik} and a_{jk}, if $Th(f_{ik}) = Th(f_{jk})$.

The following theorem is an immediate consequence of the definitions.

Theorem 9.19 In order for the system S_{ik} to be stable by the factors a_{ik} and a_{jk}, it is necessary that it admits the replacement of functions.

9.7 Conclusions

The theorems from [1, Chap. 3, Sect. 3.2, pp. 44–47] can be reformulated for levels of a quasi-fractal system in the same way as Theorems 9.9–9.13.

The open question here is the question about interchangeability of quasi-fractal system levels during its functioning.

References

1. Serdyukova, N., Serdyukov, V.: Algebraic Formalization of Smart Systems. Theory and Practice, Smart Innovation, Systems and Technologies, SIST, vol. 91, Springer Nature, Switzerland (2018)
2. Serdyukov, V.I., Serdyukova, N.A., Shishkina, S.I.: Increase in products uptime by using elements of artificial intelligence, Herald Bauman Moscow State Tech. Univ. Mech. Eng. **1**, 62–72 (2017)
3. Dvoeglazov, D.M.: The survivability and stability of enterprises of complex structure under the influence of external risks, management algorithms and risk adaptation models. Internet J. Sci. **7**(1), 1–16 (2015), Home Page http://naukovedenie.ru/PDF/72TVN115
4. Doran, G.T.: There's a SMART way to write management's goals and objectives. Manage. Rev. **70**(11), 35–36 (1981)
5. Home Page https://en.wikipedia.org/wiki/SWOT_analysis
6. Gurel, E., Merba, U.: TAT, SWOT ANALYSIS: a theoretical review. J. Int. Social Res. **10**(51), 994–1006. Home Page www.sosyalarastirmalar.com
7. Serdyukova, N.A., Serdyukov, V.I, Neustroev, S.S., Shishkina, S.I..: Assessing the reliability of the automated knowledge control results. In: 2019 IEEE Global Engineering Conference (EDUCON) (2019)
8. Suprunenko, D.A.: Matrix Groups, Science, Moscow (1972) (in Russian)
9. Bratus, A.S., Novozhilov, A.S., Rodina, E.V.: Discrete Dynamical Systems and Models in Ecology. Moscow State University of Railway Engineering, Moscow (2005). in Russian
10. Home Page https://ideafix.name/UNIVERSITY/ASU/lectures

Chapter 10
Interpretation of Solvable Groups in the Theory of Smart Systems

Abstract In this chapter, we consider smart systems whose models of determining factors are solvable groups. Also, quasi-fractal systems which modeled by quasi-fractal groups of determining factors, the levels of which are solvable groups and simple groups are considered. In Chap. 9, we have considered sustainability issues in relation to systems whose models are quasi-fractal algebraic systems. In this chapter:

- we shall introduce and consider the concept of structural sustainability of a closed associative system with a feedback,
- we shall show that closed innovation systems with commuting determining factors do not have the property of structural stability,
- a closed associative system S with a feedback modeled by a free non-abelian group is structurally sustainable with a minimum regulation range,
- theorems about tensor estimates of structurally sustainable groups.

Keywords Solvable group · Simple group · Quasi-fractal system

10.1 Introduction. Main Definitions and Finite Groups Classification

Definition 10.1 [1] A group G is called solvable if it has a subnormal series whose factor groups (quotient groups) are all abelian, that is G is a solvable group if it has a finite set of nested normal subgroups:

$$G \trianglerighteq G_1 \trianglerighteq G_2 \trianglerighteq \cdots \trianglerighteq G_n = E$$

such that G_{j-1} is normal in G_j and G_j/G_{j-1} is an abelian group, for $j = 1, 2, \ldots, n$.

Or equivalently, if its derived series, the descending normal series

$$G \trianglerighteq G^{(1)} \trianglerighteq G^{(2)} \trianglerighteq \ldots,$$

© Springer Nature Switzerland AG 2021
N. A. Serdyukova and V. I. Serdyukov, *Algebraic Identification of Smart Systems*, Intelligent Systems Reference Library 191,
https://doi.org/10.1007/978-3-030-54470-6_10

where every subgroup is the commutator subgroup of the previous one, eventually reaches the trivial subgroup of G. These two definitions are equivalent, since for every group H and every normal subgroup N of H, the quotient H/N is abelian if and only if N includes $H^{(1)}$. The least n such that $G^{(n)}$ is called the derived length of the solvable group G.

Let's reminded some necessary facts about solvable and finite groups, [1]. For finite groups, an equivalent definition is that a solvable group is a group with a composition series all of whose factors are cyclic groups of prime order. This is equivalent because a finite group has finite composition length, and every simple abelian group is cyclic of prime order. The Jordan–Hölder theorem guarantees that if one composition series has this property, then all composition series will have this property as well. For the Galois group of a polynomial, these cyclic groups correspond to nth roots (radicals) over some field. The equivalence does not necessarily hold for infinite groups: for example, since every nontrivial subgroup of the group Z of integers under addition is isomorphic to Z itself, it has no composition series, but the normal series $\{0, Z\}$, with its only factor group isomorphic to Z, proves that it is in fact solvable.

All abelian groups are trivially solvable—a subnormal series being given by just the group itself and the trivial group. But non-abelian groups may or may not be solvable.

More generally, all nilpotent groups are solvable. In particular, finite p—groups are solvable, as all finite p—groups are nilpotent. A small example of a solvable, non-nilpotent group is the symmetric group S_3. In fact, as the smallest simple non-abelian group is A_5, (the alternating group of degree 5) it follows that every group with order less than 60 is solvable. The group S_5 is not solvable—it has a composition series $\{E, A_5, S_5\}$ (and the Jordan–Hölder theorem states that every other composition series is equivalent to that one), giving factor groups isomorphic to A_5, and Z_2; and A_5 is not abelian. Generalizing this argument, coupled with the fact that A_n is a normal, maximal, non-abelian simple subgroup of S_n for n, we see that S_n is not solvable for $n > 4$. This is a key step in the proof that for every $n > 4$ there are polynomials of degree n which are not solvable by radicals (Abel–Ruffini theorem). This property is also used in complexity theory in the proof of Barrington's theorem.

The celebrated Feith–Thompson theorem states that every finite group of odd order is solvable. In particular this implies that if a finite group is simple, it is either a prime cyclic or of even order.

Any finite group whose p—Sylow subgroups are cyclic is a semidirect product of two cyclic groups, in particular solvable. Such groups are called Z—groups.

Solvability is closed under a number of operations.

If G is solvable, and H is a subgroup of G, then H is solvable, [1]. If G is solvable, and there is a homomorphism from G onto H, then H is solvable; equivalently (by the first isomorphism theorem), if G is solvable, and N is a normal subgroup of G, then G/N is solvable, [2]. The previous properties can be expanded into the following "three for the price of two" property: G is solvable if and only if both N and G/N are solvable. In particular, if G and H are solvable, the direct product $G \times H$ is solvable. Solvability is closed under group extension:

If H and G/H are solvable, then so is G; in particular, if N and H are solvable, their semidirect product is also solvable. It is also closed under wreath product:

If G and H are solvable, and X is a G-set, then the wreath product of G and H with respect to X is also solvable. For any positive integer N, the solvable groups of derived length at most N form a subvariety of the variety of groups, as they are closed under the taking of homomorphic images, subalgebras, and (direct) products. The direct product of a sequence of solvable groups with unbounded derived length is not solvable, so the class of all solvable groups is not a variety.

Burnside's theorem states that if G is a <u>finite group</u> of order $p^a q^b$ where p and q are prime numbers, and a and b are non-negative integers, then G is solvable, [1].

10.2 Representation of a Solvable Group as a Quasi-fractal. Risks of Regulation of Systems Which Models of Determining Factors are Represented as Solvable Groups

Let's represent solvable group G in a form of a quasi-fractal. A group G is solvable if it has a subnormal series whose factor groups (quotient groups) are all abelian, that is it has a finite set of nested normal subgroups:

$$G = G_0 \geqq G_1 \geqq G_2 \geqq \cdots \geqq G_n = E$$

such that G_{j-1} is normal in G_j and G_j/G_{j-1} is an abelian group, for $j = 1, 2, \ldots, n$

Let $g_j \in G_{j-1}\backslash G_j$, let $g_j = G'_j$, where $G'_j = \left(G'_j, *, ^{-1}, e\right)$ is a group isomorphic G_j, $j = 1, \ldots, n$. If $g \in G\backslash\{g_j | j = 1, \ldots n\}$ one can represent g as a single group $\langle\{e\}, *, ^{-1}, e\rangle$. So,

$\left(\{g_j | j = 1, \ldots n\} \cup G\backslash\{g_j | j = 1, \ldots n\}, *, ^{-1}, e\right)= A_1 = \langle A_1, \Omega_1\rangle$ is a quasi-fractal.

Now let's concern [3, Chap. 3, Sect. 3.2] From this it follows that what are contained in the kernel $Kerf$, do not differ from each other. The system is identified by the set of properties contained in the kernel $Kerf$: if one disables any property from $Kerf$ the system loses the entire unique set of properties from $Kerf$.

Theorem 10.2 If the system S is modeled by a solvable group of factors G_S, then S has a finite number of nested subsystems $H_k, k = 1, \ldots, n,$

$$H_S = H \geqq H_1 \geqq H_2 \geqq \cdots \geqq H_m = E, m < n,$$

such that if with probability $p_i, i = 1, \ldots, m$, the factor $g_i \in G_{i-1}\backslash G_i, i = 1, \ldots, m$, ceases to function, then the subsystems modeled by $H_k, k = 1, \ldots, n$, continue to function without changes. At the same time, G_S continue to function without changes with probability $\prod_{i=1}^n (1 - p_i)$. If $p_i, i = 1, \ldots, n$, satisfy the binomial distribution law, then maximum probability $\max(p(g))$ of the failure of the subsystem S achieves for the subsystem of the system S, modeled by $G_{m(g)}$ and

$$\max(p(g)) = p(g = m(g)) = \frac{n!}{(n - m(g))!m(g)!} p^{m(g)}(1 - p)^{n-m(g)} =$$

$$\begin{cases} \frac{n!}{(n-[(n+1)p])![(n+1)p]!} p^{[(n+1)p]}(1 - p)^{n-[(n+1)p]}, \ if \ m(g) = [(n + 1)p] \\ \frac{n!}{(n-([(n+1)p]-1))!([(n+1)p]-1)!} p^{([(n+1)p]-1)}(1 - p)^{n-([(n+1)p]-1)}, \\ if \ m(g) = [(n + 1)p] - 1 \end{cases}$$

Remark One may consider $f_{i-1} : G_{i-1} \rightarrow G_{i-1}/G_i \rightarrow 0$, and then f_{i-1} can be considered as a function of the subsystem G_{i-1} of the system G.

Proof To prove the theorem, we shall use the technic from [3, Chap. 3, Sect. 3.2]. Let G_S be a solvable group. Let's consider a finite set of nested normal subgroups:

$$G_S = G \trianglerighteq G_1 \trianglerighteq G_2 \trianglerighteq \cdots \trianglerighteq G_n = E,$$

such that G_{j-1} is normal in G_j and G_j/G_{j-1} is an abelian group, for $j = 1, 2, \ldots, n$.

Let's consider the subgroup H of the group G, which is generated by the set $G\backslash\{g\}, g \in G, H = \langle G\backslash\{g\}\rangle$, and factor $g \in G$ ceased to function with the probability p. If $H = \langle G\backslash\{g\}\rangle = E$, then the process stopes and $m = 1$. If $H = \langle G\backslash\{g\}\rangle \neq E$ then we can continue. Let's then consider the subgroup H_{j-1} of the group G_{j-1}, which is generated by the set $G_{j-1}\backslash\{g_{j-1}\}, g \in G$, $H_{j-1} = \langle G_{j-1}\backslash\{g_{j-1}\}\rangle$, and factor $g_{j-1} \in G_{j-1}$ ceased to function with the probability p_{j-1}, for $j = 1, 2, \ldots, n$. If $H_{j-1} = \langle G_{j-1}\backslash\{g_{j-1}\}\rangle = E$, then the process stopes and $m = j - 1$. If $H_{j-1} = \langle G_{j-1}\backslash\{g_{j-1}\}\rangle \neq E$, then we can continue. So, we have a chain of nested normal subgroups

$$H_{S'} = H \trianglerighteq H_1 \trianglerighteq H_2 \trianglerighteq \cdots \trianglerighteq H_m = E,$$

where S' is a subsystem of a system S modeled by $H_{S'} = H$.

Let's consider the subgroup $H = \langle G\backslash\{g\}\rangle$ of the group G, generated by $G\backslash\{g\}$ (Fig. 10.1).

So, $g \notin H$ and H continues to function.

....

$g_{j-1} \notin H_{j-1}$, so H_{j-1} continues to function for $j = 1, 2, \ldots,$.

Let's $p_0 = p$, p_{j-1}, for $j = 1, 2, \ldots, n$, satisfy the binomial distribution law. Now, let's recall the binomial distribution law (Table 10.1).

For binomial distribution law expected value of g is equal to $\bar{g} = np$, variance of g is equal to $\sigma^2(g) = npq$.

Now let's count the maximal possible value of $p(g)$.

Mode $m(g)$ of g is equal to $[(n + 1)p]$ or $[(n + 1)p] - 1$, so

	e	g	h	...	$g^{-1} \notin H$	...	$g^{-1}h \notin H$
e	e	g	h		$g^{-1} \notin H$		$g^{-1}h \notin H$
g	g	g	$gh \notin H$		$e \in H$		$h \in H$
h	h		$h^2 \in H$				
...	...	...	...		...		...
$g^{-1} \notin H$	$e \in H$	$g^{-1}h \notin H$					
...							
$g^{-1}h \notin H$	$g^{-1}hg \in H$	$g^{-1}h^2 \notin H$	...		$g^{-1}hg^{-1}$	...	

Fig. 10.1 A fragment of Cayley table of a group G

Table 10.1 Binomial distribution law

g	g_1	...	g_{j-1}	$g_j = \left\langle G'_j, *, ^{-1}, e \right\rangle$	...	g_n
	1	...	$j-1$	Number of event implementations g_j	...	n
p	p_1	...	p_{j-1}	$p_j = \frac{n!}{(n-j)!j!}p^j(1-p)^{n-j}$	...	p_n

$$\max(p(g)) = p(g = m(g)) = \frac{n!}{(n-m(g))!m(g)!}p^{m(g)}(1-p)^{n-m(g)} =$$

$$\begin{cases} \frac{n!}{(n-[(n+1)p])![(n+1)p]!}p^{[(n+1)p]}(1-p)^{n-[(n+1)p]}, & if\ m(g) = [(n+1)p] \\ \frac{n!}{(n-([(n+1)p]-1))!([(n+1)p]-1)!}p^{([(n+1)p]-1)}(1-p)^{n-([(n+1)p]-1)}, \\ \qquad if\ m(g) = [(n+1)p] - 1 \end{cases}$$

10.2.1 Theorems on Simple Finite Groups' Classification

Simple finite groups are "elementary bricks" from which any finite group can be constructed, just as any natural number can be decomposed into a product of primes. The Jordan-Hölder theorem is a more accurate way of expressing this fact for finite groups. However, a significant difference from factorization of integers is that such "bricks" will not define the group uniquely, since there may be many non-isomorphic groups with the same composition series.

The theorem is considered proven in a series of papers by approximately 100 authors, published mostly from 1955 to 2004 and containing a total of thousands of pages of text. Richard Lyons, Ronald Solomon, and (previously) Daniel Gorenstein are gradually publishing a simplified and revised version of the proof.

The classification theorem finds application in many areas of mathematics, since questions about the structure of finite groups (and their effects on other mathematical objects) can sometimes be reduced to questions about finite simple groups. Thanks to the classification theorem, such questions can sometimes be answered by checking every family of simple groups and every sporadic group, [3].

Theorem (Classification Theorem)

Any finite simple group is isomorphic to either one of the 26 sporadic groups, or belongs to one of the following three families:

1. cyclic group Z_p of the simple order;
2. alternating groups A_n of permutations of at least of 5 elements;
3. simple groups of Lie type, namely:

   classical Lie groups over a finite field, namely, Chevalley groups $PSL(n, F_q)$, $PSU(n, F_q)$, $PSO(n, F_q)$;
   and exceptional and twisted forms of Lie type groups (including the Tits group). Let's remined the following theorems, [2].

Theorem, Burnside

If G is a finite group of the order $p^a q^b$ where p and q are prime numbers, and a and b are non-negative integers, then G is solvable.

Theorem, Feith–Thompson

A finite group of odd order is solvable.

Theorem, Hall

If every maximal subgroup of a finite group G has a prime number or the square of a prime number as its index, then G is solvable

Theorem, Wielandt

If a finite group G contains three solvable subgroups with pairwise coprime indices, then G is solvable.

10.3   Structurally Sustainable Groups

10.3.1   The Physical Meaning of the Concept of System Structural Sustainability

Let a closed associative system S be modeled by a group of determining factors G_S, and a group of factors be solvable. Acting on the system with the help of regulatory functions that are described (formalized) with the help of homomorphisms, in a finite number of steps we bring the system S to a subsystem consisting of a unit factor (neutral factor). The neutral factor of the system, interacting with other factors of the system, does not have any effect on them. Thus, after a finite number of steps, the system reaches a state of equilibrium and ceases to function. Another interpretation of this fact: after a finite number of influences on a closed associative system, modeled using a solvable group of determining factors, the system ceases to function, or breaks.

If the group G_S is not a solvable one, this cannot be done. Through any finite number of control steps, a non-unit subgroup of the group G_S remains, which defines a subsystem of the system S containing a non-unit factor. The interpretation of this runs as follows. A closed associative system modeled by a group of determining factors and which is not a solvable one does not come into equilibrium as a result of any finite number of regulation steps and continues to function. Another interpretation of this fact: after a finite number of actions on a closed associative system, modeled by a group of determining factors, which is not a solvable one, the system does not stop functioning, or does not break. Let's call such a system structurally sustainable one. Let's explain the designation. After any finite number of steps of regulation of the system S, the lattice of the last subgroup remaining as a result of regulation is in some set $M_{\{f\}} = \{L(s)_i | i \in I\}$, finite or infinite. The latter depends on the properties of the group G_S. The set $M_{\{f\}}$ is called the regulation range of the system S. It is already easy to trace the analogy with the definition of stability according to Lyapunov from here

So, it is easy to obtain the following theorems in which there are examples of structurally sustainable systems.

Example 1 is presented in Theorems 10.3 and 10.4

Theorem 10.3 A closed associative system S, with feedback and commuting factors, i.e., modeled by the abelian group of determining factors G_S, is structurally unsustainable.

Theorem 10.4 A closed associative system S with a feedback modeled by a free non-abelian group is structurally sustainable with a minimum regulation range.

Proof See Definition 10.1. From Definition 10.1 and the fact that a commutator of a free non-abelian group is a free group of no more than countable rank one can get a statement of the theorem.

For example 2 we need some definitions and theorems from [4].

Definition 10.5 The equation

$$w(x_1, \ldots, x_n; g_1, \ldots, g_m) = 1 \tag{10.1}$$

of n variables with coefficients in the group G is determined by the element w of the free product $G * F(x_1, \ldots, x_n), g_1, \ldots, g_m \in G$.

If G is a subgroup of K, then under the solution of the Eq. (10.1) in K we mean a tuple $(a_1, \ldots, a_n)$ of elements of K
such that

$$w(a_1, \ldots, a_n; g_1, \ldots, g_m) = 1$$

in K that is, w goes into unity under the homomorphism $G * F(x_1, \ldots, x_n) \to K$, which maps x_i into a_i and identical on G. If $K = G$, then one speaks about the solutions of the Eq. (10.1) in G.

Definition 10.6 A group G is called algebraically closed if every finite system of equations compatible over G already has a solution in G. A group G is called strongly algebraically closed if every system of equations is a compatible over G in the case if every its finite subsystem is a compatible over G.

According to [4] the widest class of groups over which an arbitrary equation is solvable was found in 1980 by S. D. Brodsky. These are such groups, each non-unit finitely generated subgroup of which has a homomorphism onto an infinite cyclic group. These groups include free groups, torsion-free groups with one relation, many classes of solvable groups, many classes of linear groups, etc.

In 1980, S. D. Brodsky established the existence of a non-identity group, each equation over which is solvable in itself. Each group is a subgroup of some algebraically closed group.

Further, we shall try to trace whether, in the case of algebraically closed groups, there is an analogy with the Abelian group's purity theory and the theorem on distinguishing an Abelian algebraically compact group as a direct summand from the group containing it as a pure subgroup and what is the difference between these. To do this, we need a generalization of the of purity concept to the case of non-abelian groups, [5]. Here from we have the following statement.

Statement 1
An algebraically closed subgroup H of G does not have to be a retract of G
But if H is a pure subgroup of a group G, then the situation changes.

Definition 10.7 A subgroup A of a group G is called a pure subgroup of G, if every system (not only finite systems, as in [4, 5]) of equations over A, that has a solution in G, has a solution in A too.

Theorem 10.8 Let $E \to A \xrightarrow{\varepsilon} G$ be a pure embedding of a group A в into a group G, where E is a unit group, and A and G be arbitrary groups. If A is a strongly algebraically closed group, then εA is a retract of G.

Proof Recall that a group A is a retract of a group G if G can be presented as a semidirect product of a normal subgroup N and a subgroup A, that is $G = N \rtimes A$.

So, a subgroup A of a group G is a retract of G if and only if there exists an endomorphism $\rho : G \to A$ called a retraction such that $\rho \circ \rho = \rho$.

Let's consider the pure embedding

$$E \to A \xrightarrow{\varepsilon} G,$$

where A is an algebraically closed group and a diagram

$$
\begin{array}{ccc}
E \to A & \xrightarrow{\varepsilon} & G \\
& \varepsilon \searrow \;\; \nearrow' \rho & \\
& G &
\end{array}
$$

Let's define $\rho(\varepsilon(a)) = \varepsilon(a)$ for every $a \in A$. If $g, g_1, \ldots, g_m \in G \backslash A$, then let's consider all systems of relations

$$w_i(x_1, \ldots, x_n; g, g_1, \ldots, g_m) = 1, i \in I \tag{10.2}$$

over A, and

then let's consider all systems of equations over A (not only finite systems, as in [4])

$$w_i(x_1, \ldots, x_n; y, g_1, \ldots, g_m) = 1, i \in I \tag{10.3}$$

having decision in G. As A is strongly algebraically closed then the system (10.3) has a solution $y = b \in A$ in A. So, we define $\rho(g) = b$.

To complete the proof, it remains for us to show that the class of strongly algebraic groups is not empty, that is, to show the existence of a class of strictly algebraically closed groups.

In fact, according to [5], we have:

Variant of the Compactness Theorem connected with ultraproducts, [6].

Let Σ be a set of sentences of the language $\mathcal{L}$, $I = S_\omega(\Sigma)$ be a set of all finite subsets of the set Σ, and let $\mathfrak{A}_i, i \in I$, be a model of the set of sentences of i. Then there exists an ultrafilter D over the set I, such that the ultraproduct $\prod_D \mathfrak{A}_i$ is a model for Σ.

Definition 10.9 Let G be a group. Let's consider all finite systems of equations over G. Let $w_i(x_1, \ldots, x_n; y, g_1, \ldots, g_m) = 1, i \in I$, be all systems of equations of n variables with coefficients in the group G over G having decision in G.

Let's embed G into strongly algebraically closed group which can be wrote in the form $\prod_D \mathfrak{A}_i$ according to Variant of the Compactness Theorem connected with ultraproducts, [4], where $I = S_\omega(\Sigma)$ is a set of all finite subsets of the set a set of all finite subsets of the set

$w_i(x_1, \ldots, x_n; y, g_1, \ldots, g_m) = 1, i \in I$. Then $\prod_D \mathfrak{A}_i$ is called a strongly algebraically closed envelope of the group G and is designated by $SA(G)$. It is easy to see that G is embeddable into $SA(G)$.

Theorem 10.10 A closed associative system S with a feedback modeled by a strongly algebraically closed group is structurally sustainable with a minimum regulation range.

Proof From the fact that algebraically closed groups are simple one gets that strongly algebraically closed groups are simple.

The following notion from [7] helps to connect the notion of algebraically closed groups and strongly algebraically closed groups.

Definition 10.11 [7] A group A is Noetherian by equations if every system of equations from $x_1, \ldots, x_n$ with coefficients from A is equivalent to some of its finite subsystem over A for any natural number n.

The Noetherian condition for equations is equivalent to the Noetherian condition of the Zariski topology defined on the set A_n, when algebraic sets are taken as the prebase of a system of closed sets, i.e., sets of solutions of systems of equations over A.

Also, [8], every group represented by matrices over a commutative Noetherian unitary ring is Noetherian by the equations. Here from, one gets, in particular, that free non abelian groups are Noetherian by the equations.

It was established in [8, 9] that the class of equationally Noetherian groups together with each group contains its finite extension and the quotient group in the normal subgroup closed in the Zariski topology.

One more very interesting theorem from [9] runs as follows.

Theorem 10.12 Every finitely generated group with a solvable word equality problem is embeddable in any algebraically closed group.

Here from one can get the following theorem.

Theorem 10.13 Any closed associative system S with a feedback can be embedded into an infinite structurally sustainable system.

Proof According to [5], and Theorem 10.9, each group is a subgroup of some algebraically closed group. Besides it, algebraically closed groups are simple and cannot be generated by finitely many elements, as B. Neumann proved.

Open Questions

1. Here from the question about the sustainability of innovative and pseudo-innovative systems arises.
2. Here from the second question about the P—sustainability of innovative and pseudo-innovative systems arises, where P is some property.

10.4   Tensor Estimation on Smart Systems Modelling by Finite Simple and Finite Solvable Groups

In this section we shall consider tensor estimations on smart systems modelling by finite simple and finite solvable groups. Let's use Definition 9.1. from the Chap. 9, Sect. 9.2:

Let Q be some quality of a system S, which can be estimated with a help of a set

$$H_Q(S) = \{h_1, h_2, \ldots, h_n\}$$

of numerical characteristics. We shall assume, that these numerical characteristics are functions depending of the time: $h_i = h_i(t), t = 1, \ldots, n$.

We shall construct a tensor estimate of the Q—quality of the functioning of the system S as a homomorphism of a group of factors G_S, determining the system S into a group $GL(n, R)$ of linear homogeneous transformations of the vector space R^n. We shall use the same construction as in [3].

We represent the tensor estimate as a homomorphism

$G_S \rightarrow GL(n, R)$, where G_S is a group of factors defining the system S, $GL(n, R)$ is a group of linear homogeneous transformations of a vector space R^n, n—a number of quantitative indicators that assess the education management system.

Let's assume that G_S is a finite group, consisting of n elements: $G_S = \{a_1, a_2, \ldots, a_n\}$ and $h_i = h_i(t) = h_i(a_i, t)$ is a numerical characteristic of a factor a_i.

So, a homomorphism $f : G_S \rightarrow GL(n, R)$ can be considered as the tensor estimation of the quality Q.

Definition 9.1 A homomorphism $f : G_S \rightarrow GL(n, R)$ can be considered as the tensor estimation of the quality Q.

10.4.1   Tensor Estimates of Structurally Sustainable Groups

Let's remined that:

- free non abelian groups are Noetherian by the equations,
- A beautiful linear representation of the group F_2 was found in 1947 by Sanov [4]. **Theorem.** The correspondence $x_1 \mapsto \begin{pmatrix} 1 & 2 \\ 0 & 1 \end{pmatrix}$, $x_2 \mapsto \begin{pmatrix} 1 & 0 \\ 2 & 1 \end{pmatrix}$, defines an isomorphic embedding of the group F_2 into $GL_2(Z)$. If one replaces the written matrices onto $A_\mu \mapsto \begin{pmatrix} 1 & \mu \\ 0 & 1 \end{pmatrix}$, $B_\mu \mapsto \begin{pmatrix} 1 & 0 \\ \mu & 1 \end{pmatrix}$, then the corresponding representation is an isomorphism for all complex μ with the condition $|\mu| \geq 2$.

Theorem 10.14 (on tensor estimation of a closed associative system with a feedback)
Let S be a closed associative system with a feedback, which is modeled by a group of determining factors G_S. If there exist an epimorphisms $\varphi_3 : F_2 \to G_S$, where F_2 is a free non abelian group of a rank 2, (that is, for example, a group G_S has two generating elements, for example system S has only two determining factors) and an epimorphism $\varphi_4 : GL(n, R) \to G_S$, then tensor estimation

$$h : G_S \to GL(n, R)$$

of the system S (according to a model G_S) coincides with the restriction of tensor estimation of a strongly algebraically closed group by the set G_S:

$$f : SA(G_S) \to GL(n, R)$$

that is $h = f \restriction G_S$. So, if a closed associative system with a feedback is modeled by such a group of determining factors G_S, that there exists epimorphisms of a free non abelian free group of a rank 2 onto G_S and of a $GL(n, R)$ onto G_S, then the tensor estimation of a system S is a restriction of a tensor estimation of its supersystem modeled by a strongly algebraically closed group.

Proof Let's embed a group G_S of determining factors of the system S into a (strongly) algebraically closed group. This can be done by Theorem 10.9. Let's consider the chain of embeddings:

$$F_\omega \to [F_2, F_2] \to F_2 \to GL(2, Z) \to GL(2, R) \to GL(n, R)$$

where $F_\omega \to [F_2, F_2]$ is an isomorphism of a free non abelian group of a countable rank onto a commutant of a free non abelian free group of a rank 2. Let

$F_2 \to GL(2, Z)$ be an embedding which is defined by Sanov's theorem, $GL(2, Z) \to GL(2, R)$ be a natural embedding, $GL_2(R) \to GL(n, R)$ be a natural embedding.
Let's consider the following diagram

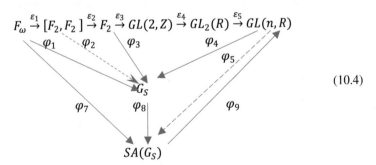

$$(10.4)$$

where $SA(G_S)$ is a strongly algebraically closed group, containing G_S as a subgroup. Let $\varphi_2 : [F_2, F_2] \to G_S$ be an epimorphism, let

$\varphi_1 : F_\omega \to G_S, \varphi_1 = \varphi_2 \varepsilon_1, \varphi_3 : F_2 \to G_S, \varphi_4 : GL(n, R), \varphi_5 : GL(n, R)) \to SA(G_S), \varphi_6 : SA(G_S) \to GL(n, R), \varphi_7 : F_\omega \to SA(G_S), \varphi_8 : G_S \to SA(G_S), \varphi_9 : SA(G_S) \to GL(n, R), \varphi_3 = \varepsilon_3 \varepsilon_4 \varepsilon_5 \varphi_4, \varphi_5 = \varphi_8 \varphi_4$, and φ_9 is a retract.

From the Definition 9.1 one gets that φ_9 is a tensor estimation of strongly algebraically closed group $SA(G_S)$. $\varphi_9 \varphi_8$ is the tensor estimation of the group of factors G_S.

Theorem 10.15 (on tensor estimation of structurally sustainable systems with a minimum regulation range)

Let S be a closed associative system with a feedback, which is modeled by a group of determining factors G_S and $\varphi_1 : F_\omega \to G_S$ be an epimorphism of a free nonabelian group of countable rank F_ω onto G_S. If there exists an epimorphism

$$\varphi_2 : GL(n, R) \to G_S$$

then tensor estimation $h : G_S \to GL(n, R)$ of the system S (according to a model G_S) coincides with the restriction of tensor estimation of a strongly algebraically closed group by the set G_S:

$$f : SA(G_S) \to GL(n, R)$$

that is $h = f \mid G_S$. So, if a closed associative system with a feedback is modeled by such a group of determining factors G_S, that there exists epimorphisms of $GL(n, R)$ onto G_S, then the tensor estimation of a system S is a restriction of a tensor estimation of its supersystem modeled by a strongly algebraically closed group.

Proof Let's embed a group G_S of determining factors of the system S into a (strongly) algebraically closed group. This can be done by Theorem 10.9. Let's consider the chain of embeddings:

$$F_\omega \to GL(2, Z) \to GL(2, R) \to GL(n, R)$$

and the following diagram:

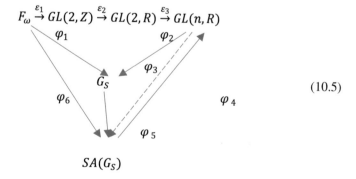

$$(10.5)$$

where $SA(G_S)$ is a strongly algebraically closed group, containing G_S as a subgroup, $\varphi_1 : F_\omega \to G_S$, $\varphi_2 : GL(n, R) \to G_S$, $\varphi_3 : GL(n, R) \to SA(G_S)$, $\varphi_5 : SA(G_S) \to GL(n, R)$, $\varphi_6 : F_\omega \to SA(G_S)$, $\varphi_4 : G_S \to SA(G_S)$, $\varphi_3 = \varphi_4\varphi_2$, $\varphi_6 = \varphi_4\varphi_1$ and φ_5 is a retract.

From the Definition 9.1 one gets that φ_9 is a tensor estimation of strongly algebraically closed group $SA(G_S)$. $\varphi_5\varphi_4$ is the tensor estimation of the group of factors G_S.

In what follows, one can use the theorems on the epimorphic images of the group $GL(n, R)$, and about epimorphic images having two generators, of the group $GL(n, R)$

References

1. Home Page https://en.wikipedia.org/w/index.php?title=Special:ElectronPdf&page=Solvable+group&action=show-download-screen
2. Home Page https://lektsii.org/13-61547.html
3. Serdyukova, N., Serdyukov, V.: Algebraic Formalization of Smart Systems. Theory and Practice, Smart Innovation, Systems and Technologies, SIST, vol. 91. Springer Nature, Switzerland (2018)
4. Olshansky, AYu., Shmelkin, A.L.: Infinite groups, Itogi Nauki i Tekhniki. Ser. Lying. prob. mat. Fundam. Directions **37**, 5–113 (1989)
5. Mazhuga, A. M.: Verbally Closed Subgroups, Thesis for the Degree of Candidate of Physical and Mathematical Sciences. Moscow State University, Moscow (2018) (in Russian)
6. Keysler, H.J., Chang, C.C.: Model Theory, North–Holland Publishing Company. American Elsevier Publishing Company Inc, Amsterdam, London, New York (1973)
7. Gupta, C.K., Romanovsky, N.S.: Noternity by some equations solvable groups. Algebra Logic **46**(1), 46–59 (2007)
8. Baumslag, G., Myasnikov, A., Remeslennikov, V.: Algebraic geometry over groups. I. Algebraic sets and ideal theory. J. Algebra **219**(1), 16–79 (1999)
9. Baumslag, G., Myasnikov, A., Roman'kov, V.: Two theorems about equationally noetherian groups. J. Algebra **194**(2), 654–664 (1997)
10. Home Page https://ru.wikipedia.org/wiki/Classificationofsimple_end_groups
11. Reyman, A.G., Semenov-Tian-Shansky, M.A.: Group-theoretical methods in the theory of finite-dimensional integrable systems. Dyn. Syst. **VII**, 116–225 (2003)

Printed in the United States
by Baker & Taylor Publisher Services